THE OUTBACK
vs
THE WILD WEST

THE OUTBACK vs THE WILD WEST

VOLUME 2 OF
THE WILD WEST IN AUSTRALIA AND AMERICA

Jack Drake

Central Queensland
UNIVERSITY
PRESS

Copyright 2006 Jack Drake

This book is copyright. Apart from any fair dealing for the purpose of private study, research, criticism or review, as permitted under the Copyright Act of 1968, no part may be reproduced by any process without written permission. Enquiries should be made to the publisher.

First published in 2006 by Central Queensland University Press

Distributed by:
CQU Press
PO Box 1615
Rockhampton Queensland 4700
Ph: (07) 4923 2520
Fax: (07) 4923 2525
Email: cqupress@cqu.edu.au
www.outbackbooks.com

National Library of Australia
Cataloguing-in-Publication data:

Drake, Jack.
The Wild West in Australia and America : An Historical Comparison of True Frontier Tales

Bibliography.

ISBN 1 876780 66 5 (volume 1)
ISBN 1 876780 67 3 (volume 2)

1. Frontier and pioneer life - Australia
2. Pioneers - Australia - Biography
3. Frontier and pioneer life - United States - West
4. Pioneers - United States - Biography
5. Australia - History
6. West (U.S.) - History

I. Title

909.8

Cover design and typesetting by Jane Dorrington

Front cover painting: 'Cobb and Co. Down Cunningham's Gap' by the late Hugh Sawrey. Courtesy of his wife Mrs Gill Sawrey, Benalla, Victoria 3672

Back cover painting by Alan Clive Grosvenor

Printed and bound by Watson Ferguson & Co.
Moorooka, Brisbane

This book is dedicated to my dear wife, Stellamary Matheson Drake

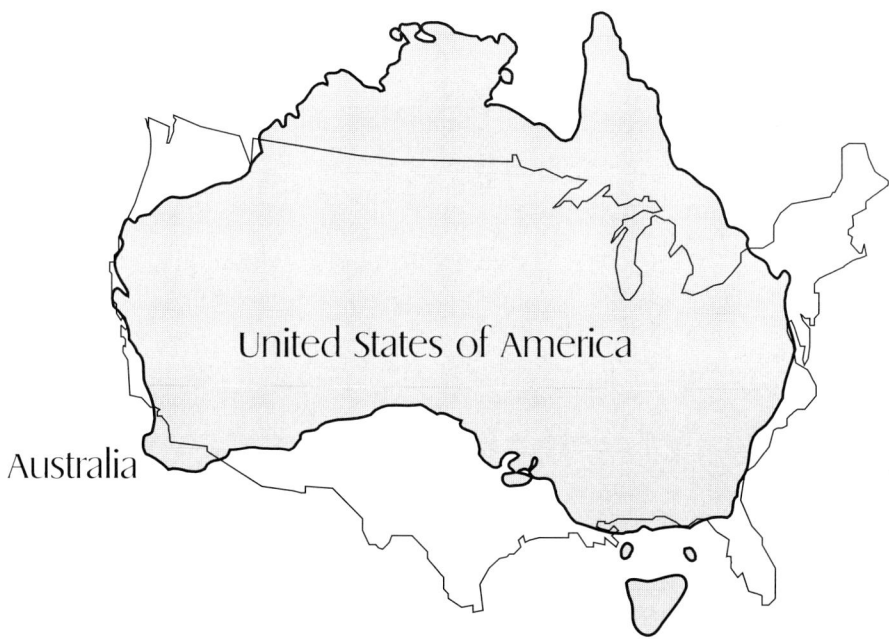

Comparison between the size of the continent of Australia and mainland United States of America

Area of Australia: 7,687,000 sq.kms

Area of mainland United States of America: 7,829,000 sq.kms

Acknowledgements

Many people have given me a lot of help in the preparation of this book. First and foremost, thank you to my wife Stella who has given a large part of 2004 to preparation and compilation. It would not have happened without her, and thanks also to Tammy, Vicki, Leon, Carney and Jeffrey, our children, who have all given support and interest to the project.

Of all people involved, special thanks must go to Jeff Simpson of Severnlea and John Osborne of Greymare, for their assistance in presenting a very real portrayal of the horsemen of The West. These blokes are 'fair dinkum' and both have also been wonderful research sources.

My gratitude is also due to Jim McJannett of Thursday Island, Frank Uhr of Brisbane and Peter Poole of Warwick who gave unstintingly of their wide knowledge and historic material to assist a writer making his first attempt at a factual work.

Di Rieger and her staff at the Stanthorpe Shire Library have been marvellous. Nothing was too much trouble for them and they have helped me so much in tracking down rare and obscure writings from two continents.

My good friend Barney Belford has been involved from the start and has given me information on a host of subjects as well as being a rapid and accurate proof reader.

My thanks must also go to Angus Frame, Caloundra; James and Anna Tyson, Mt. Tamborine; Chris Sarra and the staff and pupils of Cherbourg State School; Bruce Simpson, Caboolture; the Queensland Police Museum, Brisbane; Col Newsome, Glen Innes; Jill Bowen, Sydney; Bert and Fay Paull, Brisbane; Neville and Joyce Bryant, Stanthorpe; Peter Hiendorff, Brisbane; Ian Tinney, Crows Nest; Bill Gray, Finlay; Ned Winter, Cecil Plains; Geoff Frame, Chinchilla; Barry Cox, Knoxfield; Mary Hodgeson, Mooloolah; The Historical Society of Beaudesert; Terry Underwood, *Riveren*; Kaye Bleakly and Janet Keith PRORODEO, HOF Colorado Springs, USA; Bill Wade, Mundubbera; Otago Early Settlers Museum, New Zealand; Gary Fogarty, Millmerran; Denise Bender, Camp Hill; Neil McArthur, Ballarat; Bob Dellar, Stanthorpe; Tom and Louise Oliver, Sarina; Trevor Hollitt, Glen Lyon; Merv Buckly; The Australian Geographic Society; Graham Murchie, Ballandean; John Skinner, Warwick, Ian McLaine, Brisbane; Jean Green, Katanning; Trevor Wren, Severnlea; Paddy Cannon, Atherton; Jaqueline and Keith Belford, Eukey; Caval

Graeme, Eukey; Mr. And Mrs. D.Duncan-Kemp, Oakey; The Stockmans Hall of Fame, Longreach; The American Consulate, Sydney; Peter and Sue Ingall, Glen Aplin, Brendan and Sylvia Raleigh, Wandoan; Ted Robl; Cloncurry and District Historical and Museum Society Inc.

All these people and organisations have generously supplied material and expertise.

Finally I owe a big thank you to Colin Munro of the A.B.C. for writing the Foreword and to my Publisher, Old Silvertail, Prof. David Myers and his excellent off-sider Jane Dorrington.

Foreword

I was delighted to be invited to write a foreword to Jack Drake's superhuman effort with the two tomes comparing the Australian and American Wild Wests.

The old expression Mrs Kelly wouldn't let Ned play with whomever, applies very much to Jack Drake. Ned was determined but I think on this occasion Jack has out gunned him.

When Jack wrote and asked me to read his "scribblings" I had no idea that two Brisbane telephone book sized manuscripts would thunder into my mailbox at home in New South Wales enclosing a typical Jack Drake epistle. I quote "Well mate, here it is. A year's worth of hitting the books and a life's worth of interest in the subject. We've got some marvellous stories out there and I haven't even scratched the surface." Heaven help us when Jack's about ten feet down.

You don't tangle with Jack Drake lightly. Six feet of bearded bushman, tossed off more buckers than the rest of us have had hot dinners; a wrestler with the 'boghi' on the shearing board; gone in hips and knees; a naturalised Australian from across the Tasman; and above all you don't poke muck with strong minded dwellers from the Queensland Granite Belt or highly accomplished Bush Balladeers.

I am lost in admiration at the wide loop Jack has cast in his researches. No skimpy references to history, rather carefully cross referenced material that cannot be denied.

This is no sixpenny western; rather a litany of gripping facts, humour, high days and low of human beings in Australia and the United States.

I don't knock the Stars & Stripes but I am proud that this grand literary effort from Jack Drake was written under the Southern Cross.

Well done Jack! Thanks for the great ride through two nations' outback histories.

Go kindly old mate!

Colin Munro
Manager Regional Liaison
Australian Broadcasting Corporation

Foreword

Contents

Acknowledgements vii

Foreword ix

Introduction xiii

Drovers and Trail Drivers 1

Australia: Joseph Hawdon - Australia's first long distance drover • Sutherlands' trip with sheep • Nat Buchanan • The greatest drove ever – 3,220 km with 20,000 head • Buchanan achievements • The longest trip • The world record dry stage • South to north with sheep. • Perils of Droving • Reflections of a drover • Women of the stock routes • End of an era

America: America's early cattle drives • Nelson Story's trail drive to Montana • River crossings • The greatest stock swim • A bad dry trip

Duffers and Rustlers 32

Australia: Song of the sheep stealer • Harry Redford • Kimberley Cattle Duffer - Jim Campbell • The Keane Tyson brand affair • Henry Beecham and the thoroughbred brumbies • Galloping Jones – the ultimate larrikin • Duffers of today.

America: The Tombstone cowboys • John Chisum and the legal rustle

Black, Red and White Wars 49

Beginnings and duration in both countries.

Australia: Australian conflict • Pemulwuy and the Eora resistance • The war for the Bathurst Plains • The Tasmanian annihilation • Myall Creek • Bielbah - the freedom fighter • Fraser - the avenger • The fighting Kalkadoons • A few short accounts of Aboriginal resistance with white reaction • The 29 year war • Milbong and Dundalli – the scourges of early Brisbane • Warriors of the Lockyer • Zac Skyring's saviours • *Conniston* • Aborigines on horseback

America: Early disputes • The Battle of Beechers Island • Custer's last stand • Custer's conquerors • Sitting Bull • Crazy Horse • The avengers • Geronimo and the last wild Indians

White Against White 87

The difference between two continents.

Australia • The battle of Vinegar Hill • Eureka Stockade • The Queensland Shearers' wars • The strike of 1891 • The 1894 strikes • The siege of *Dagworth*. • The aftermath • Other Australian conflicts

America: The Mountain Meadows massacre • The Johnson County war • The Tonto Basin war • Sheep versus cows

Women of the Wild West 116

Australia: Frontier females • The lady pirate of Old Sydney Town • The Eulo Queen • Australia's Annie Oakley – the Claudie Lakeland story • Lola Montez • Mrs. Black Jack Reed of Booroloola • Cockney Fan • The lady bushranger

America: Dora Hand – diva of Dodge City • Belle Starr – the bandit queen

Stagecoach Whips, Pony Express Riders, Bullockys and Mule Skinners 142

Similar beginnings. Freighting on two continents.

Australia: Cobb and Co – Australia's Wells Fargo • Australia's master reinsmen • Australia's Pony Express • The humble packhorse • Packhorse mailmen • The Fizzer • Bureaucrats don't understand flooded rivers • 'Speargrass' Jack Gard – Gulf mailman • Days of the Dromedary • Riverboats

America: John Butterfield and the beginnings of Wells Fargo • Russell Majors and Wadell • Freighting giants and coaching wannabees • Ben Holladay – the stagecoach king • America's best whips • Black Bart – state of the art stagecoach robber

Horseborne Endurance Feats 165

Buffalo Bill's and West Fraser's rides.

Australia: When fever hits the doctor rides • Billy Mateer's race against the flood • The extraordinary journey of the Stringybark Fox

America: How Louis Remme outrode bankruptcy • From the Arctic to the Tropics • The Sheridan to Galena ride

Buckjumpers and Roughriders 181

Australia: The reason for the legend • Lance Skuthorpe • Legendary buckers • Saddle development. Dargin's Grey and Breaker Morant • *Rocky Ned* • *Curio* • Who was the best • Non competitive champs • Roughriding ladies • Today's champions

America: Booger Red • The bruising of Buster Ivory • Steamboat and others • *Midnight* • *Five Minutes to Midnight* • The Breaker's American counterpart • A roughriding artist and writer • Mind over matter

The Western Myth 210

Australia: Australia's Crooked Mick of the *Speewah* • Mysterious monsters • Myth –v- Reality

America: Western superheroes in America • Mysterious monsters

Conclusion 219

Endnotes 220

INTRODUCTION

If you have just picked up this book, the beginning of the story is contained in *The Wild West in Australia and America Volume 1* CQUP, 2005.

Volume 1 tells of the earliest white adventurers who penetrated two continents. The whalers, sealers, castaways and convict bolters of Australia, and the fur traders who first moved among America's Indians. It deals with notable clashes that occurred when lawless frontiers were governed by the rule of the gun and the mounted police, sheriffs and marshals who pursued killers and brought law to the Outback of Australia, and the western lands of America.

Australian bushrangers and American outlaws rode vast regions claimed by cattle kings like Australia's James Tyson and the United States' Col. Charlie Goodnight. The stockmen and cowboys who were the life-blood of the huge stations and ranches, feature as do buffalo hunters from the great plains of North America and Australia's tropical north.

Now the story continues………..

DROVERS & TRAIL DRIVERS

Store cattle from Nelanjie. Their breath is on the breeze.
You hear them tread a thousand head in bluegrass to the knees.
The lead is on the netting fence. The wings are spreading wide.
The lame and laggard scarcely move – so slow the drovers ride.
But let them stay and feed today for sake of Auld Lang Syne.
They'll never get a chance like this below the border line
And if they tread our frontage down, what's that to me and you?
What's ours to fare, by God they'll share! For we've been droving too.

Will Ogilvie

Droving in Australia and America

The most publicised stock drovers in the world were the American cowboys who brought herds of longhorn steers up the Chisholm and other western cattle trails in the wake of the Civil War.

The era of the 'Great Cattle Drives' has been lauded through story, song, T.V. series and Hollywood epics until the terms 'cowboy' and 'trail herd' have been applied to almost any movement of stock under their own power anywhere.

Australia, however, had a history of large-scale droving that began in 1836 and continued until the mid 1960s.

The number of cattle and sheep that walked to market, railhead, fattening and breeding properties in Australia, far exceed those from America's trail driving era, which lasted for under thirty years.

American drives were traditionally comprised of larger herds because the bulk of watering was done at rivers and creeks where the cattle could spread out along the banks. An average trail herd was from two to two and a half thousand head with about fourteen men in attendance.

An American trail crew was made up of a trail boss, cook who drove the chuck wagon and nine or ten cowboys with the herd who rode in the positions known as point, swing, flank and drag. Point riders were the most experienced men who rode either side of the lead steers while the trail boss rode ahead. Swing and flank riders held the sides together and the least experienced drag riders brought the back of the herd along. The crew was completed by a wrangler who minded the horses in the day time and a nighthawk who cared for them at night. Cowboys shared the night watch duties. There were two men with the cattle at almost all times.

Australian road mobs were between a thousand and fifteen hundred, a convenient number to water at holes, dams and troughs to which they would be brought up in 'cuts' of two hundred and fifty to four hundred at a time.

A droving team was comprised of the drover in charge or boss drover, horse tailer, cook and two or three ringers with the cattle. Positions the men took in Australia are known as lead for the rider in front, wing for the ringers who held the sides together and the back was known as the tail.

Some drovers used wagonettes to cart the gear and tucker but the need to find grass meant most long trips with store bullocks, were done with packhorses. The first watch was usually taken by the horse-tailer while the others had their evening meal, and the others took turns through the night. If the cattle were touchy or a storm was likely, two men would double watch, but under normal circumstances the cattle were held on the night camp by one ringer.

Australian Droving Plant & Road Mob

4 men with cattle in emergency + Horse tailer makes 5
Drover-in-charge has no set position but works where he feels there might be a problem

American Outfit & Trail Herd

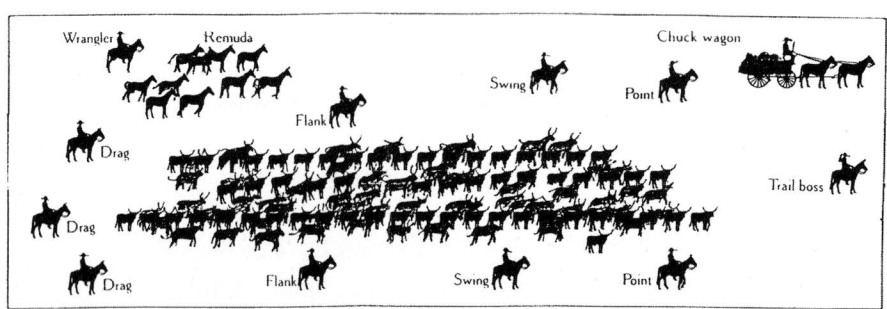

Night watchers or night guards as they were known in America, always made a noise as they rode around the cattle. This let the stock know where the man was at all times and had nothing to do with 'singing the cattle to sleep' as many people believe. Anyone with a good voice was certainly welcome on the road but most cowboys and ringers couldn't carry a tune in a bucket.

Joseph Hawdon, Australia's First Long Distance Drover

Large-scale droving was well established in Australia long before the days of the Chisholm and other American cattle trails with the first recorded trip of any magnitude happening in 1836. Smaller trips had taken place before then but when Joseph Hawdon and his partners, John Gardiner and Captain Hepburn, took 300 cattle from the Murrumbidgee River to Port Phillip, (Melbourne), it set the stage for the Australian droving epic. Hawdon took up the first cattle run on the Yarra River and built a hut and yards on the site later occupied by Scotch College.[1]

Joseph Hawdon obviously had a liking for overlanding ventures being the first man to hold the Melbourne to Yass mail contract. He commenced the service on 1st January 1838 and accompanied his contractor, John Bourke, on the first trip with riding and packhorses. They completed the 300 plus kilometres trip in three and a half days. The mail went on to Sydney by wheeled vehicle. Hawdon was back at his *Howlong* run near Albury a few days later to begin his most famous droving feat.

On 13th January 1838 Hawdon was under way again. He left Howlong and made his way to Adelaide, a distance of 1,200 kilometres following the Murray River most of the way.[2] The trip took twelve weeks to complete and traversed country where later that same year, seven members of the Faithful party were killed by Pangerang warriors of the Ovens River country. The Pangerang were a warlike tribe and George Faithful made another attempt to occupy the Ovens River country only to be sent packing for the second time, but Joseph Hawdon managed to traverse their tribal lands with no contact.

Hawdon became a well-known pastoralist owning several properties in Victoria before returning to England in 1840. He returned to Australia in 1856 and two years later sold out and moved to New Zealand where he established three sheep stations in the foothills of the Southern Alps.

Sutherland's Trip With Sheep

Not only cattle made long treks across inland Australia and one of the epics was George Sutherland's journey inland from Rockhampton, Queensland in search of land to settle.[3] George and his brother John followed the Queensland coast north in 1863 with 8,000 merinos in search of sheep country. They stopped to shear the mob at Suttor Creek when spear grass seeds began accumulating in the wool, carted the wool to Mackay for shipment, and allowed the ewes to lamb. When the newborns were big enough to travel, they pushed on towards Townsville before turning inland up the Cape River.

After crossing the Dividing Range they followed the Flinders watershed into the Hughenden area to find all the good country taken up. The Sutherland brothers pushed westward still chasing the Flinders course out past the Richmond district, but all of the huge plains of that region had been claimed for grazing.

In places where the Sutherlands could have claimed squatter's rights, the grass was so long and rank, it was unsuitable for sheep, and in one forty mile section between the Leichhardt and Gregory Rivers, logs had to be dragged crosswise behind the drays to beat down a path for the sheep.

They held their westerly direction crossing the watersheds of the Gulf Country rivers until they reached the Gregory River near the Northern Territory border without finding suitable country. Determined to establish a sheep property, George Sutherland sent a party to report on the Barkly Tableland discovered by William Landsborough, and when they returned with a favourable report, the brothers pushed on through the rough, hilly country to the west.

Stalked by hostile Aborigines, and at times desperate for water and grass, they finally came out into open Mitchell Grass plains. A thirty-five mile dry stage led them to Lake Mary a large permanent waterhole on the Georgina River north of where Camooweal now stands.

The seven months, 2,100 kilometres trip had a rather amusing finale when the perishing sheep smelt the water in Lake Mary. The Sutherlands had been driving through the night after a hole at the head of Chester Creek proved dry, and they reached the lake around midnight. A large number of Aborigines, who had never seen a sheep in their lives, were camped on the shore. In the way of animals desperate for water, the 8,000 sheep behaved as if nothing was in their way and put the tribesmen and their families completely to rout by charging straight over the camp in their frenzied rush for the shoreline.

The Aborigines were understandably terrified by these strange creatures who ran over them in the dead of night, and a famous corroboree from the Georgina River country near Camooweal had its theme based on Sutherlands' sheep and their rush for water.

The Sutherland brothers settled at Lake Mary and named their new station *Rocklands*.

Nat Buchanan

There is no doubt that Australia's and the world's greatest drover was Nathaniel (Nat) Buchanan.[4] Some vast movements of stock happened in ancient times like the 20,000 mules and 5,000 camels used by Alexander the Great to carry off booty from the sack of Persepolis in 331 BC, but in more modern, droving circles, Buchanan was King. Nat or Bluey as he came to be known, was born in Wexford, Ireland in 1826. He arrived in Australia with his family aboard the *'Statesman'* in January 1837. The family took up land in the New England region of northern New South Wales in 1839 where Nat grew up in a world of stock and horses.

Always adventurous, Nat and two of his brothers sailed to California with a large dose of gold fever to join the Forty-niners in the fabulous gold rush that began there in the tailrace at Sutter's Mill.

After working their passage home on a windjammer having failed miserably on the gold fields, the Buchanan brothers were more than happy to return to pastoral pursuits.

With a nickname like 'Bluey' it is natural that most Australians make the assumption Nat was a redhead. In reality he was a tall, spare framed man with brownish grey hair and the far seeing gaze of a pathfinder. A popular legend has him riding the farthest reaches of Australia holding a parasol over himself but this happened only in later years when the ravages of a life in the open had caused some skin problems.

Nat began overlanding cattle to South Australia and the Victorian gold fields and became known as a reliable drover with a flair for organization. In 1859 he made his way north towards Rockhampton where he rode into the camp of some other New England emigrants, the Gordons near Ban Ban where the family intended to settle. The meeting would come to have a profound effect on Nat Buchanan. Catherine Gordon was to become his wife and her brothers Willie, Hugh and Wattie became lifetime friends and business partners.

In Rockhampton Nat fell in with explorer William Landsborough. He travelled to the north west with Landsborough on his first exploring trip into Queensland's Outback. He and Landsborough took up *Bowen Downs* and *Mt.Cornish* Stations south of Hughenden in Queensland, which would later become the scene of Harry Redford's famous cattle duffing feat (chapter 2). Four years later low prices forced them to walk out.

Nat did some droving and managing. He then returned to the New England region for a few years taking up a selection near Kempsey, trying tin mining at Bendemeer and running brumbies on the Tablelands. Then the lure of the Outback drew him back to Queensland.

In April 1878 Nat began droving 1,200 cattle from Aramac in Queensland 2,250 kilometres to *Glencoe* on the Daly River south of Darwin, or Palmerston as it was known then, in the Northern Territory for *Glencoe's* owners Travers and Gibson. This was his first trip across the Old Gulf Track pioneered by Ludwig Leichhardt the explorer in 1844-45, and first traversed with stock by Wentworth D'arcy Uhr who took cattle across in 1872 to supply the Overland Telegraph, and horses as breeding stock for Matthew Dillon Cox. (chapter 3 vol.1).

Nat's trip had its fair share of drama with Aboriginal attacks and supplies running dangerously low causing Nat and his brother-in-law Wattie Gordon to take packhorses ahead 480 kilometres to Katherine for fresh supplies. Hugh Gordon minded the cattle with the other men, and one, named W. Travers, a nephew of one of the cattle owners, was tomahawked to death by Aborigines while alone in camp.

Cattle as well as a few horses, were speared by natives. Others were taken by crocodiles. Nat and Wattie arrived back with Greenhide Sam Croaker, whom they had met in Katherine, on the very day that Hugh made the decision to abandon the cattle and try to get himself and his men to safety.

The delays caused the party to be caught by the wet season and once again they had to camp the cattle near Mt.Minn. A few old sheets of iron from the abandoned Overland Telegraph Base Camp at least provided a dry galley to cook in and keep a fire going.

Two hundred calves were born while they waited for the weather to break and once again Nat had to go for rations, this time to *Daly Waters* with Sam Croaker as offsider. Two months after their return with fresh supplies, the weather broke enough to travel on to Elsey Creek where rain once again stopped progress. Hugh Gordon was struck down with malaria there but luckily the Elsey Telegraph operator had a supply of quinine and he recovered well.

Nat had to go ahead to Katherine once more for food and it was the end of May 1879 before the first large mob to travel across the top of Australia reached its destination at *Glencoe*.

The Greatest Drove Ever - 3,220 kms with 20,000 head

Nat Buchanan's greatest droving feat began two years after his first crossing of the Old Coast Track when C.B.Fisher and Maurice Lyons were negotiating to take *Glencoe* over from Travers and Gibson as well as securing leases for *Marrakai* and *Daly Rivers* Stations. Fisher and Lyons had also obtained vast leases in the Victoria River district and they offered Nat the job of taking 20,000 breeding stock over the Gulf track to their Territory properties. The modern world's largest, long distance movement of stock was planned and Nat Buchanan was to be the drover in charge.

Cattle were assembled from C.B.Fisher's properties in Queensland. The biggest part of them were put together in the south and Nat assigned Willie Gordon, to take charge of the southern mobs while he went north to arrange the northern contingent. Drays, wagons, camp gear, harness, saddlery and provisions all had to be procured. Draught and saddle horses were bought or supplied from Fisher's stations.

What a scene it must have been at the main departure camp near St.George in Southern Queensland. Pounding hooves and swirling dust as stockmen broke in new horses and took the sting out of others freshly mustered to ready them for the long journey north. Huge mobs of bawling cattle driven in from the stations and held on open camps while skilled riders on beautifully trained campdraft horses cut out road cattle from the station mobs. Dust clouds were roiled over the main mobs and the cuts, as order emerged out of seeming confusion and 16,000 breeding stock were readied for the long trek north.

Willie Gordon started the mobs of 1,200 each north at the correct spacings to allow sufficient distance so the mob of each individual drover did not interfere with those in front or behind. Cattle in mobs of more than 1,200 are very hard to water effectively in Australian conditions where often relatively small watering points provide the only drink available. Larger mobs are unwieldy to handle and experience has shown that 1,000 to 1,500 at most is a convenient road mob.

Nat was there to count the cattle off on their eight miles a day stages and before the dust of the last mob had settled, he was on his way north by coach, train and ship to Townsville to ready the northern operation.

Willie Gordon rode up and down the line of seven mobs that strung out over seventy miles. Until the cattle settled into the routine of travelling, some wild scenes were enacted. A young stockman, Paddy Fitzpatrick, was killed when a mob rushed off a night camp and he was trampled beneath the churning hooves when his horse fell trying to swing the leaders.

Willie nearly came to grief during another rush. While he was camping with one of the mobs the cattle took fright. He flung himself across one of the night horses and galloped out to help turn the lead. In the confusion Willie and the drover in charge, Harry Farquharson, went up one side of the mob and an inexperienced Aboriginal stockman called Spider tried to turn them from the other. This created a perilous situation known as a 'chinaman's lane' in droving circles with men trying to turn the leaders into a ring from both sides. Luckily the melee sorted itself out but next day the drovers did a lot of riding to muster lost cattle. Spider learned a valuable lesson and from then on he always advertised his presence on night watch by "Singem strong fella corroboree all right" in his words.

The southern mobs worked their way north to their rendezvous at the Flinders River for which five months had been allowed while Nat was busy organising horses and cattle in the north. He went straight to *Richmond Downs* to make arrangements for the 4,000 cattle there to be mustered and drafted, then began co-ordinating horses, plant and supplies and getting them to the meeting point in time to meet the road mobs from the south.

The northern cattle were broken into four mobs and allocated to droving teams with Hugh and Wattie Gordon in charge of two of them. Right on time the southern cattle arrived and the *Richmond Downs* mobs strung off down the Flinders in the wake of the now seasoned travellers from St.George.

The last big load of supplies from eastern settlements was freighted down from Normanton 192 kilometres north on the Gulf of Carpentaria. Burketown to the west lay only thirty two kilometres off the stock-route but it was not a major supply point and could not be relied on for major amounts of provisions.

Late monsoon rains halted cattle and the Normanton supply wagons. The centre mobs were stranded for a while below the point where the Flinders and Cloncurry Rivers merge in a maze of overflow channels. The cavalcade of wagons, drays, cattle, horses and men struggled through the low lying scrub and swamp area with its hordes of mosquitoes between *Canobie* on the Cloncurry and *Floraville* on the Leichhardt Rivers and there Nat's woes were further compounded when his drovers staged a strike for extra wages.

It is not clear how many men were involved but they certainly had Buchanan over a barrel and their demands were agreed to. Some men decided to pull out regardless, and this left the teams shorthanded meaning extra night watches and less sleep for everyone left.

It was apparently an unwritten law in America that a man employed with travelling stock stayed on to finish the drive. Things were different here and many stockmen left or joined en route. When store mobs started they were usually pretty wild, but once they settled down, cost-conscious boss drovers often welcomed the chance to shed someone from the team. On this drive though, Nat needed everyone he could get.

More difficulties arose when some horses went missing and Hugh Gordon, an expert bushman and tracker, had to hand his mob to a replacement while he went after them. Hugh found a dozen or more including some close hobbled and hidden in an out of the way place. Horse stealing was something of a growth industry in the Gulf Country in those days, but Hugh Gordon was a hard man to fool.

The wagons from Normanton arrived as the lead mobs reached Burketown. Nat had engaged extra men there to travel down with the teamsters expecting some of the party to pull out before the uninhabited country.

Normal length watches and stomachs with a few tinned or dried delicacies in them, after the dreadful sameness of salted beef and damper day after day, restored morale among the troops. Some easy travelling on good, well-grassed country gave the animals a chance to build up a little before they took on the Gulf Track and the hundreds of miles of wilderness it represented. This route is shown as No.6 in the Principal Stock Routes of Australia map following.

Nat rode ahead to the leading mobs and saw them start the Gulf Track proper at Turnoff Lagoon on the Nicholson River. A few of the fainter hearts pulled out there, no doubt daunted by distance, fever, unfriendly natives, crocodiles and all the rest of the perils of the country they were about to enter. The real bushmen, however, grinned wryly and turned their horses' heads west. One can only wonder if those resolute men would have agreed with the poet Banjo Paterson who sprung to prominence four years later, when he penned the line '*A drover's life has pleasures that the townsfolk never know.*'

West of Turnoff Lagoon, the country became scrubby clay soils with poor unnutritious grass. The heat was oppressive and on the sour coastal country cattle, and particularly horses, lost condition. The men were reduced to baking Johnny cakes for a few selected riding horses to give them a bit more heart and skirmishes with Aborigines became common as they entered The Calvert River country whose tribes were notoriously hostile towards intruders white or black.

Hemmed in by coastal mangrove swamps and rugged ranges, the drovers were in constant danger of attack. No doubt the presence of large mobs of cattle day after day diminished the natives' food supply and they compensated by spearing cattle and even horses for meat.

Scurvy was a constant threat to bushmen with a diet lacking in Vitamin C but Nat made sure he carried lime juice to ward off the disease known as 'Barcoo Rot'. The only other death on the trip, besides Paddy Fitzpatrick, was a man named Sayle who died of malaria or gulf fever as it was known, in spite of quinine being carried.

They crossed better country around the Robinson River and a couple of rest days were taken near the McArthur River before the final and possibly the worst obstacle. More scrub and sand country before plunging into the Gutta Percha Scrubs to the west of Limmin River.

At last the hungry, weary stock came out into well grassed country around Roper Bar and from there the hundred and thirty odd kilometres through to the *Elsey* was a drover's dream. Open plains with tree lined creeks and tons of feed to build the stock up for the final leg.

The first mob came into Katherine with Nat in attendance in early October 1882 and after that the last two hundred kilometres to *Glencoe* were simplified by following the Overland Telegraph line. Nat stayed at Katherine until all the mobs were safely across the river then he rode ahead to supervise the arrival at *Glencoe*.

Three other mobs had followed Nat's caravan of cattle also belonging to Fisher and Lyons and bound for *Glencoe*. One of them was in the charge of Jack Warby a very competitive individual who determined to be first to arrive. Warby had driven his cattle hard since Roper Bar and trying to ford the Katherine River in a hurry. They began milling mid-stream resulting in about seventy drowned beasts and a very lucky escape for a drover whose horse was drowned when caught in a group of swimming cattle. Needless to say, Warby did not win the race.

The mobs were counted and dispatched to their various blocks as they walked into *Glencoe* and once the men were paid and the droving plants disbanded, Nat returned to Sydney by sea. His 18 months, 3,220 kilometres odyssey earned him the title beyond any doubt of the world's greatest drover.

An unhappy footnote to the greatest droving feat of all time was that *Glencoe* was unsuccessful. It was eventually abandoned due mainly to redwater fever caused by ticks that came into the Territory with water buffalo imported from Asia.

Nat Buchanan's Achievments

Nat Buchanan's droving and exploring feats are the stuff of legend.[5] He pioneered the Murranji Track from *Top Springs* to *Newcastle Waters* in the Northern Territory, which cut hundreds of miles off the coast route, causing the closure of the old

Coast road. Nat also made the first east west crossing of the Barkley Tableland in company with Greenhide Sam Croaker.

The author at Nat Buchanan's grave, Walcha Cemetery, New England Tablelands.

Photo by Stella Drake
Courtesy Jack Drake collection

He established *Wave Hill* Station, explored deep into the Tanami Desert and took cattle from *Flora Valley*, the property of his brothers-in-law, the Gordons, south to Beringara in Western Australia, pioneering a stock-route to the Murchison Gold Fields. He pioneered the export of live cattle to Singapore from Derby in the Kimberley. Nat explored the dry heart of Australia in an attempt to find a practical stock-route to the west and as an old man of seventy he tried again to find a route west using camels supplied by the South Australian Government.

Nat Buchanan retired to *Kenmuir,* a small, lucerne farm in the New England country of his boyhood. In his outback years, he took up many leases and worked *Bowen Downs* and *Wave Hill* but somehow fortune eluded him. Perhaps he was more concerned with discovering than having and holding, and whenever he was contained in one place his restless eyes were ever drawn to the horizon.

He passed on in September 1901 and now lies beside his beloved wife Catherine, in the cemetery at Walcha. If you ever visit that small town on the New England Tableland, take a few moments to stand in reverence of the greatest drover the world has ever seen. It is a moving experience.

The Longest Trip

The longest trip in Australia with stock began in March 1883 near Goulburn, New South Wales. Donald McDonald, a Scot from the Isle of Skye, had settled with his parents on a property called *Clifford Downs* in the Goulburn area during the 1820s.[6]

Donald heard of land in huge quantities in the Kimberley region of Western Australia after the explorer Alexander Forrest passed through the region.

After a trip by sea to Derby by one of his sons to inspect the country inland, the McDonald clan saw the chance to own a kingdom and plans for the journey were made. Shortly before departure Donald McDonald was killed in a riding accident but his sons Will and Charlie made the decision to go anyway.

Reports vary on the number of cattle that left Goulburn but it was somewhere between 500 and 1,000 plus forty horses and two bullock teams accompanied by the McDonald brothers, their cousin Donald McKenzie, three stockmen and a Chinese cook.

1883 was a bad drought year in Queensland and the brothers had to go miles out of their way to find grass and water. Forced to camp on the Thomson River because further travel was impossible, their entire mob of cattle perished when the waters dried up. Donald McKenzie and the three stockmen pulled out but the McDonald boys were fighters not prepared to let their father's dream die.

Willie and Charlie took jobs in the area, and by taking their pay in cattle, they built up another mob. Their Chinese cook stuck with them and it is a shame and a sign of those times that history has not bothered to record his name.

Under way once again, the brothers battled through to Burketown and plunged into the scrubs of the Old Gulf Track. Aboriginal attacks, stock losses and all the pitfalls of that hostile road to the west dogged their footsteps and at the Roper Bar Charlie McDonald fell so ill with fever he had to be taken to the coast by a party of prospectors who were fortunately camped in the area.

One of the wagons collapsed totally at the Roper and loading the remainder of their stores on the other, Willie and his loyal Chinese companion faced the remaining wilderness. Willie McDonald's trip from Roper Bar to *Fossil Downs* near the Fitzroy River in Western Australia would sound too far-fetched even for Hollywood.

Each day he would scout the day's travel, return to set his cook on the road with the wagon and bring the cattle along. Part way he managed to engage one or two men from another party (once again records differ on the number) and just when things looked a bit better, their stores were rifled by natives. This made ships biscuits and salt beef the bill of fare.

The Durack and Costello families' trek from *Thylungra* in Western Queensland to the Ord River was under way in the same years of the McDonald migration. When a gaunt, ragged, fever-wasted Willie McDonald came to their camp one night after following their tracks up, Long Michael Durack almost shot him by mistake, taking him for a hostile Aborigine.[7] Willie had tied his horse up away from camp not wishing to disturb the sleepers. He had long worn out his last pair of boots and the ragged barefoot apparition that appeared in the moonlight could well have taken a bullet had he not spoken before Michael fired.

Three and a half years after leaving Goulburn Willie McDonald reached the junction of the Fitzroy and Margaret Rivers. Three hundred and seventy cattle and fourteen horses completed the 5,630 kilometres journey. *Fossil Downs,* the property Willie and his brother established, is still held by the McDonald family today.

When Willie drove his battered wagon over to Derby for supplies it became the first vehicle to cross the Australian continent. The McDonald brothers' historic trek became the longest droving trip in Australia's and possibly the world's history.

The World's Record Dry Stage

Inverway Station in the Victoria River district, was taken up by three brothers Harry, Archie and Hugh Farquharson, after being located by Nat Buchanan and his son Gordon.

The brothers took the lease on the 6,220 square kilometre station, worked to stock it and moved in during 1894 with a mob of cattle they had overlanded from Inverell in northern New South Wales. They were true pioneering stock, born in the New England district. Seeing a chance in an unsettled area, the brothers made their way to the Top End. Harry was a boss drover for his uncle Nat Buchanan on the epic drove across the Gulf Track with 20,000 head.

The Farquharsons were of hard-working Scottish descent and they all pulled together to build *Inverway* by droving and doing other work to get the place off the ground. Harry was the stockman, Hughie the booker-keeper, Archie the all-round man and an exceptional bush carpenter, but they were all versatile, and could turn a hand where it was needed.[8]

1909 was a very dry year up that way after four continuous years of drought. Could they get cattle away? The weather said "No!" The budget said "Yes". It was June and the weather was cold – conditions perfect for droving, and the cattle were in good store condition. The brothers decided to have a go.

They got to Top Springs on the Armstrong River with the mob of just over 1,000 big bullocks in good shape ready to take on the notorious Murranji Track. Route 9 on the Principal Stock Routes of Australia map following. While they spelled the cattle, two drovers returning from a trip east told them they would not get a drink for the mob before *Newcastle Waters*.

There were three watering points on the Murranji in those days before government bores were installed - Murranji, Yellow Hole and The Bucket with its companion waterhole, Hickidy, a few kilometres away on the same creek. They were all low and sufficient only to give horses a drink. They were of no use for a big mob of thirsty bullocks which would trample them into mud in minutes.

The Farquharsons spelled the cattle for a day or two at Top Springs and held them back from the water until late afternoon to make them thirsty on the day they departed. They let the mob fill up with all they would drink and with the wind in their faces, six men, forty six horses and 1020 big nor' west shorthorn bullocks took on the Murranji.

Every beast there was at least four years old and dry country bred. They strung away up the jump-up where the track climbs to a tableland and plunged into the

Lancewood and Bullwaddy scrub. The brothers were taking a huge gamble on the weather. Had it turned hot or the wind changed, the situation would have been desperate and a perish would probably have resulted.

They drove by night and day putting the cattle on camp around eleven o'clock each morning to spell through the hottest hours until that was useless because the thirsty bullocks would not settle. Then they took them as steady as they could by day but bullocks, that are keen to drink, will stride out. During the nights, they made great time with the leaders following close behind a weaving bobbing pinpoint of light from a hurricane lamp carried by the man in the lead.

The horses got a drink at Yellow and Murranji Holes with the cattle being swung well out on the windward side so they would not get a smell of the water. By the time they came out of the scrubs onto open plain country on the evening of the third day, the mob was silent, but still marching. They were spread out over a bigger distance now and there was a desperate air of purpose about drovers and cattle. The horses not in use were taken ahead to Bucket hole where they got a drink and as the mob came up, the men swung the leaders south east to miss the smell of the small, muddy puddle.

Sixty tail-end bullocks were cut off and taken in to the water. These were the weakest, and what they could get might just see them across the last of that hellish track. The tail-enders got a drink of sorts, and Bucket Hole was nothing but churned up mud when the men pushed them out on the tracks of the rest.

It was late afternoon on the fourth day. A low moaning sound arose from the perishing mob and a few of the weakest bullocks were dropping out to die. Suddenly, a stir seemed to run through the leaders as they raised their noses into a wind that carried the blessed scent of water. New life surged through the mob and the cattle began to trot. They were so strung out there was no danger of a smothering rush into *Newcastle* lagoon and the leaders hit the water to drink and drink and drink.

Hours later, the last remnant staggered in and sunk their noses. The Farquharsons had done the impossible. Together with their drovers and horses, the brothers had driven over 1,000 bullocks two hundred kilometres with only a scant drink for a few weak tail enders, and lost five bullocks.[9] Luck was riding with them in the form of cool weather and wind in the bullocks' faces which always makes them walk out, but no dry droving feat in the world has ever exceeded this trek and it is a safe bet to say none ever will.

Dennis 'Jim' Ronan once did a 128 km dry stage with fat bullocks from *Welford* over the Barrier Track into South Australia in 1886.[10] This would probably stand as a record amongst fat cattle drovers whose charges would never foot it with big, aged cattle in store condition.

The Farquharson brothers held *Inverway* Station until the last surviving brother, Archie, sold the place in 1947. Archie continued to live there until he died in 1950 aged 88 and is buried on the property.

South to North With Sheep

In September 1870 Ralph Milner began putting a large mob of sheep and the men, provisions and equipment to handle them, together at Killalpaninna near Kapperamanna, North Adelaide, South Australia.

A bonus was offered by the South Australian Government, of £2,000 for the first 1,000 sheep or one hundred cattle to arrive in the Northern Territory, and Milner intended to take advantage of this offer and sell his sheep as food for workers on the Overland Telegraph line. Ralph Milner's wife had recently died and his bold plan of a south-north Australian traverse with stock was in part a commercial venture and also a way of dealing with his grief, by making a new start on a distant frontier.[11]

Nine whites and four Aborigines made up the human members of the expedition. Their charges were numerous. 7,000 sheep, 300 horses, two bullock drays with teams, one horse drawn wagon with a twelve horse team, two spring carts also horse drawn, one year's provisions, the makings for fifty pack saddles should the vehicles have to be abandoned and the whole cavalcade accompanied by a total of twenty-five dogs, ten for sheep work and fifteen grey and staghounds for hunting. The Milners were keen sportsmen.

The drovers and teamsters were made up of Ralph Milner as leader, his brother John as second in command, Edward Kirk, Jack Brown and Bill Lamb as bullock drivers, John 'Yorky' Thompson as wagon driver, Jack Wooding, Arthur Ashwin and Harry Pybus as stockmen. Two of the Aboriginal men ran away shortly after passing *The Peake* Station, which was the extent of settlement, but one man Charley, and his wife Fanny, saw the job through.

The sheep picked their way north and by the end of January, they were in stone country where they spent four days building a cueing pen and shoeing the working bullocks with half horseshoes to protect their hooves.

Heavy rain and flooding from the north at the Finke River, halted progress for five weeks and eventually the party managed to swim the sheep across this river practically in single file. They followed the Finke to its junction with the Hugh River, which they then ascended to near its source in the McDonnell Ranges.

Again they camped for six weeks while lambs were born and grew enough to travel. During this break they hunted, branded foals and spent a week fighting off a plague of rats, which inundated the camp in thousands keeping all hands busy guarding the food.

The drove got under way again and worked its way north. Near Tennant Creek 2,000 sheep were lost to poison weed and a camp was set up at Attack Creek while the country ahead was checked out. John Milner was in charge of the camp while Ralph and some others explored. John was an easy going, kind hearted man and when a Warramunga tribesman came into the camp, he welcomed him despite the custom of the time of not allowing tribal Aborigines into white campsites.

When John Milner's attention was diverted, the big warrior brought a war club down on his head then turned to throw a boomerang at Arthur Ashwin who was nearby. Ashwin fired two shots with a double-barrelled duelling pistol he was carrying. The first took the Warramunga in the stomach and the second knocked him over when he turned to run.

A kangaroo dog joined the fight grabbing the native under the arm while Ashwin ran to a horse tethered nearby for a rifle. The dog gave up and returned and fortunately for the warrior who was running wounded, Ashwin's rifle misfired.

John Milner's face was smashed between his ear and eye and both eyes dangled on his cheeks. They buried him under a big gum tree with a tin plate recording his epitaph and a heavy timber fence to protect the grave.

While in camp at Red Lily Lagoon on Christmas Day 1871, a Mr. Patterson of the Overland Telegraph came by and purchased 2,000 sheep at £2/10/- a head delivered to Bitter Springs (Mataranka) about thirty miles away. After the sheep were delivered, the Milner party camped on a ridge as the wet season was coming down with a vengeance.

For three months they lived on mutton and water, their other supplies being all used up. They put in their time building packsaddles and breaking in horses.

In March the monsoon broke and the balance of the sheep were sold again to Patterson of the Overland Telegraph for £1/17/- each. The Telegraph Company also bought wagons, teams and surplus horses and Ralph Milner and his men's 2,600 kilometres droving odyssey was finished.

Apparently governments were as two faced then as they are now. Part way through the trip and unknown to the drovers, the proposed £2,000 bonus for the first stock into the Territory, was cancelled so Ralph Milner missed out on some of the profit from his venture.

Perils of Droving

Many of Australia's and America's droving trips were marvels of courage and endurance. Some horrific incidents occurred and in some cases, the drovers themselves died or were lucky to get out with their lives.

Western movies have made great use of the 'stampede' or 'rush' as Australia knows it. Rushing cattle are the most commonly perceived danger to drovers and some tragic accidents have happened as a result of galloping mobs.

Drovers who have experienced a rush say it happens in an instant. One moment the cattle are quiet and suddenly, the night erupts with sound as the whole herd runs blindly into the darkness. Cattle as grazing animals are extremely sensitive to the mental state of their companions. The telepathic transferral of fear is one of the grass eater's main forms of defence and this is the cause of rushes. One animal's fright can spread through the mob like wildfire causing the herd to run as a unit rather than a group of individuals.

The only thing to do, is to get low along the horse's neck if there is timber around, and trust the night horse. Any attempt to pull a horse around with the reins could interfere with its progress and bring horse and rider crashing down. Once the leaders were reached the man and horse combination would attempt to turn them back and cause the mob to 'ring' in a circle to halt their frenzied flight.

Despite the obvious dangers, there have been surprisingly few fatalities in rushes. The quality of the night horses and experience of the drovers would certainly have a bearing on this. The worst smashes have occurred when mobs rushed in the direction of the camp but they would seldom go over the fire that was always situated between the cattle and the swags and equipment. The person on night watch would stop to build up the fire when it burnt down. Nevertheless, mobs have gone over camps killing men and wrecking gear.

As soon as trouble started, every man in the camp ran for the night horses tied up nearby. If the man on watch could contain the mob before they really got going and spread off camp, so much the better.

Sometimes cattle would take fright and stop in only a short distance. Australian drovers called this a 'jump' or 'splash'. When the mob ran until halted by riders, it was a 'rush' and when loss of life and/or property occurred, the incident became known as a 'smash' which was the general term for loss of animals, men and gear in the droving game.

Dry conditions have caused more smashes than any other single cause in Australia and cattle have perished from thirst in America as well. In the 1900 – 03 drought, a mob of 2,000 *Rocklands* bred Kidman bullocks failed to find water after leaving *Carcoory* Station and every head perished.[12] The same fate befell 1,000 breeding cows Kidman had en route to *Pandi Pandi* in those same desperate years.

Many a lonely forgotten grave lies on the banks of rivers in Outback Australia and up the American Cattle Trails where herdsmen were drowned when things went wrong swimming large mobs across flooded waterways.

Hostile tribesmen caused havoc to travelling stock and their herders on both continents. When the exploring Jardine brothers left Rockhampton to establish a settlement, called Somerset, near the top of Cape York Peninsula in 1864, they were dogged by hostiles all the way up the western side of the Cape.[13] After ten months and 1,900 kkilometres that included swamp, jungle and rivers, as well as crocodile and Aboriginal attacks, Frank and Alex Jardine arrived with fifty cattle from their original two hundred and fifty and their horses were reduced from forty two to twelve.

Droving hazards were not all life threatening. Strayed stock and horses, walkouts by men, cantankerous, incompetent and sometimes drunken cooks all contributed to the difficulties encountered getting a mob to its destination. Sore-footed bullocks were an ongoing problem in stony country and drovers solved this by 'cueing'. Lame animals were caught and thrown so the men could tack half horseshoes flattened at the heel, to their cloven hooves.

Principal Stock Routes of Australia

Some stock-routes used during historic trips. Major stock routes which later led to railheads. Australia was criss-crossed by designated routes — far too many to include. Early drovers followed natural waters until the network of artesian bores was established.

Alice Springs/Oodnadatta/ Port Augusta/Adelaide	No. 17	Cloncurry/Winton/ Diamantina/Birdsville	No. 15
Barkly Stock Route	14	Gulf Country to Victoria	11
Barrier track	18	Hawdon's Howlong to Adelaide 1838	8
Birdsville Track	10		
Buchanan's Stock Routes:		Katherine to *Newcastle Waters*	12
Aramac to Glencoe 1878	2	Murranji track	9
St. George to Glencoe 1881	3	*Newcastle Waters*/Alice Springs	16
Richmond to Ord 1883	4	Old Gulf Route	6
Flora Valley (Halls Ck) to Beringarra 1892	5	Ranken/Georgina/Diamantina	13
		Sutherland's Rockhampton to Rocklands 1863	7
Canning Stock Route	1		

The Overland Telegraph Line   - - - - - -

The route the Milner brothers followed from Port Augusta to the Roper River

There is reference to the practice of shoeing cattle in America as well. Beef steers from Charles Goodnight's J.A. Ranch in the Texas Panhandle walked to Dodge City and when Goodnight began upgrading his herds with durham (shorthorn) bulls, the crossbreds were not as hard footed as longhorns.[14] J.A. cowboys often shod sore footed animals to allow them to keep up with the herd along the Palo Duro – Dodge City Trail that went through some hard country.

Reflections of a Drover

The last of the cowboys who came up America's cattle trails died in the early to mid 20th century. Another thirty years will see the passing of the last of Australia's old time drovers.

The following is a brief overview of conditions in the droving game from the 1940s to the 1960s. Jeff Simpson went into Australia's northern cattle country as a young man in the wake of World War II. Jeff is the real thing. He has lived a life most of us can only read and romance about.

Here is his own description of Australian droving – the recollections of an old time drover.

> "Droving in Australia remained virtually unchanged from the early days of colonisation and expansion of the pastoral industry, until the mid 1960s. The introduction of large road transport at this time supplanted the drovers although mobs even today use the stock routes in hard times for feed.
>
> For over one hundred years droving remained an integral part of the pastoral industry for every beast at sometime, somewhere in Australia, walked to the sale yard, trucking yard or abattoir.
>
> The animals were placed in the hands of a professional drover termed 'the drover in Charge' who was usually paid a sum of money per head per one hundred miles for long trips, or an agreed sum for short trips. The drover would sign a contract with the owners of the stock after an agreed head count and price, and from that point on, he would be solely responsible for the stock until they were delivered at the contracted destination.
>
> The drover provided the horses and equipment (this was referred to as 'the plant'), and paid and fed other men required to shift the herd. Droving plants varied in size depending on length of trip, numbers in the herd being driven, or even the type of country.
>
> The size of a plant could range from a couple of men, three or four dogs, a buggy and a few horses in the case of a sheep drover travelling through fenced stock routes, to quite a large plant shifting store cattle from say, the Kimberley or Northern Territory. The larger type of

plant could be moving anywhere between 1,000 and 1,500 head for 1,000 miles or more, and take as long as eight months to complete. A typical plant for such a trip would consist of the drover, three stockmen, a cook and a horse tailer, eight packs and sixty head of horses. Some drovers used wagonettes and in the late 1940s – 50s light trucks, but the packs were still taken as back up if the trucks bogged or broke down.

The horse plant could be roughly broken up in the following categories:

8 pack horses and 2 spares

5 day horses per man

as many as 12 night horses (used only for watching the cattle on night camp in sets of twos or threes – one for the man on watch, and the rest tied up in case of trouble. These were sure-footed horses with good night vision and cattle sense.)

The drover's reputation and life depended on these horses if the cattle rushed or were playing up.

The horse tailer was responsible for packing up the camp and shifting the horses to the new night camp. The supplies were usually packed in the following manner:

1. The Cook's tucker pack, which carried cooked cold meat, damper, tea, coffee, sauce, curry powder, baking powder, tin or enamel dishes, knives, forks and spoons in other words, just about everything the cook needed to throw on a quick basic meal.

2. The meat pack. Drovers killed a beast whenever needed most of which was salted and would last two or three weeks. (The owners usually provided a number above the tally for this purpose).

3. One or two water canteen packs with four gallons capacity a side, of shaped galvanised steel which fitted snugly against the pack saddle.

4. A shoeing pack containing spare horse shoes, shoeing gear and a short piece of railway iron used as a substitute anvil to shape shoes.

5. Three dried tucker packs. These carried bags of flour, sugar, rice, salt for salting meat, potatoes, onions, dehydrated vegetables, dried fruits etc. Cooking utensils such as billies and rolled metal 'Bedourie' style camp ovens were carried strapped on top of swags which lay crosswise over the tops of the pack saddles.

Jeff Simpson filling water canteens on a packhorse from the trough at a mill on the Barkly Tableland with his brother, Bruce's horse plant in the background.

Photo courtesy Jeff Simpson collection)

The horse tailer was responsible for redistributing the loads as supplies were used, so bags were evenly balanced each day. The health and general welfare of the horses were his prime task although he was often used as an extra man with the cattle when the need arose.

These store drovers were the elite of the profession travelling vast distances along unfenced routes which could contain mountains, scrubs, deserts or flooded rivers, with semi-wild, unhandled cattle. It was from such men and epic trips, that much of the romantic aura and folk lore of the legendary drovers has arisen.

Though there were many major routes that followed rivers, there were many that crossed the more arid areas where surface water was scarce. In the early 1900s successive governments drilled and equipped bores on such routes. These bores varied between sixteen to twenty four miles apart. This gave rise to the practice of a dry day to cross the longer distances, and was done in the following manner:

The drover would hold the cattle on a bore for a couple of hours while the animals had a good drink, then mid afternoon feed them slowly out to the night camp four or five miles on. The next day with an early start just after daylight, they would feed and walk on another ten to twelve miles to the camp without a drink. Again with an early start the next morning, they would walk through to be watered by dinner time on the next bore.

In those days the cattle were always watched at night and the horse plant restrained by being hobbled with numerous horses wearing bells so their position could always be detected.

The method of putting a mob on night camp was to feed them up quietly near the fire just before dark. The drover would circle the mob bunching them together into a relatively small area adjacent to the fire. A mob of 1,250 would occupy less than an acre quite comfortably. After the drover satisfied himself that the stock were settled, he would hand over the first watch to the horse tailer who circled the mob at a slow walk while whistling, singing or reciting poetry which the cattle found reassuring.

A watch with a full team was around two hours with each man taking his turn except the cook who did the dog watch (a short watch just to allow the drover in charge to get his meal). The drover always track rode his camp if cattle had played up at night to ensure none had slipped away unnoticed.

This routine was followed by thousands of men Australia wide when things were going well on the road. Bad cattle, storms, dry stages, short handedness all altered this routine drastically. A rough diet, long hours, sleeping on the ground for weeks in a very basic swag of canvas with a couple of blankets, made this a hard and dangerous life. All in all the Australian drover was a pretty tough and skilled character who deserves his place in our history."

America's Early Cattle Drives

Following the surrender of Confederate troops at Appomatox in April 1865, the men of ranching families made their way back to cash-poor and cattle-rich Texas ready to drive cattle north to the railheads for eastern markets. Cattle had been driven to market in America long before this though. As early as 1750 large herds of cattle were running along the Mississippi River in the Natchez district northwest of New Orleans and in the Perdido, Tombigbee and Chickasaway valleys north east of Mobile in Alabama.[15] Herders on horseback were recorded driving cattle to New Orleans and Mobile by 1812 for shipment to the West Indies and Atlantic coast cities.

In 1846 Rancher Edward Piper, drove Texas cattle to Ohio and during the Californian Gold Rush, T.J.Trimmier drove 500 steers there in 1848, which sold for $100. each. W.H. Snyder spent two years driving a herd to San Francisco arriving in 1849, and in 1855 a group of Texans drove cattle all the way to New York City. Charlie Goodnight's future partner, Oliver Loving, made drives before and during the Civil War. He drove Texas cattle to Illinois in 1855 and was recognised as America's most experienced trail driver in those times. Sadly, no records were kept of these historic drives other than departure and arrival notices in newspapers.

These were America's first western trail drivers but in 1866 the era of the fabled great cattle drives began. The enterprising Texans hit a snag. Those first few herds driven north infected the cattle of the farming country they traversed with 'Texas fever'.

This was caused by ticks, but nobody knew it then. The Longhorns were immune to tick fever but the eastern stock of Kansas sickened and died in the wake of Texas herds that came north in previous years.

The drives of 1866 were a shambles when Kansas Farmers genuinely concerned for their stock and gangs of self styled vigilantes known as 'Jayhawkers' turned the herds back at the border.[16] One such confrontation made a Texas trail boss utter an order that was to become a byword for ex-Confederate Texans approaching Kansas who had supported the Union in the War between the States. "Bend 'em west boys" he snarled, "Nothing in Kansas anyway except sunshine, sunflowers and sons of bitches."

In 1867 an enterprising cattle dealer from Springfield, Illinois named Joseph Geating McCoy provided a solution to the problem by entering into a contract with the Kansas Pacific Railroad to provide a shipping point at Abilene west of the Kansas farming districts. Abilene was the first trail town but as settlement and the railway advanced, the trade shifted to Ellsworth, Newton, Wichita and Dodge City.

For all the publicity the era of the great trail drives received, it was under thirty years duration before railroads made long distance driving unnecessary. The time spent on the Chisholm and other trails was seldom over three months per herd. Australia on the other hand had a history of droving that spanned from 1838 until the mid 1960s when the golden age of droving finally gave way to the stock transporters. The average time of a trip from northern Australia with store bullocks, was over five months and there can be no doubt the most extensive movements of stock that walked to their destination happened in Australia.

Nelson Story's Trail Drive to Montana

Perhaps the most dangerous trip with cattle ever, due to attacks by hostile natives, took place in America in 1866. Nelson Story was a cool headed young man who 'struck it rich' on the Alder Gulch Goldfield in Montana.[17] Story had his fill of gold mining and with around $12,000 of the profit on his mining venture sewn into his clothing, he headed south with the idea of buying cattle in Texas for sale in Kansas.

A thousand steers at $10 a head were purchased in Texas. A crew was hired and the cattle were headed north. Story and his cowboys experienced the usual trials and tribulations of trail driving particularly flooded rivers, 1866 being a very wet year. They arrived at Baxter Springs to find the previously mentioned snarl-up of herds turned back by the Kansas border guards.

Nelson Story did some quick thinking about meat-hungry miners in the north-west and made a bold decision to take his investment all the way to Virginia City in south west Montana. Story headed his cattle west to skirt around the extent of Kansas settlement and worked his way north then east to Fort Leavenworth.

As an added avenue of profit, he bought a wagon and work oxen and loaded it with provisions saleable on the mining fields before heading westward along the Oregon Trail. The Army Officers at Fort Laramie looked grim when Story's cavalcade arrived there. The Bozeman Trail, the most feasible route to Montana, was overrun with hostile Sioux Indians who had had enough of the ceaseless swarm of immigrants and miners who were invading their country.

Despite warnings of dire consequences, Nelson Story bought rapid fire Remington breech loading rifles for all his men and pushed on into Indian Territory. As the herd approached Fort Reno a war party of Sioux attacked wounding two cowboys and running part of the herd off. Story took the wounded men to Fort Reno then followed the raiders into the hills where he and his men used their new weapons effectively enough to recover the cattle.

At Fort Phil Kearny, Colonel Carrington put Nelson and his men under arrest and forbade them to continue. They were ordered to hold the cattle several miles from the Fort so the army stock's grazing was not eaten out. Story was in an impossible situation. They were just as vulnerable to attack, as they would be travelling, so he called his men together and proposed that they slip off in the night.

One man George Dow disagreed and in case he gave the alarm, he was placed under arrest with a promise of being freed to return to the Fort as soon as the rest were clear. Once away from Colonel Carrington's authority, Dow was frightened to return alone, and chose to remain with the herd. Story and his men drove the stock by night and fed them under guard during the day.

Twice more the Indians attacked but the Remington repeaters proved a far greater deterrent than the Army's single shot Springfields. Out of the twenty-seven men in the party only one life was lost.

On 9th December 1866 Nelson Story's herd and supply wagon rolled into Virginia City, Montana where he disposed of his well travelled stock and provisions at boom prices.

The Bozeman Trail closed to all traffic for four years after Story's herd came through. The war that raged through that part of Montana resulted in an Indian victory second only to Custer's last stand. The brash young Captain William J.Fetterman, who had boasted that with eighty men he would ride through the entire Sioux nation, had his command wiped out to a man. The decoy party that lured Fetterman to his death was led by a young warrior of the Ogalalla Sioux named Curly who was soon to become known and feared as Tashunka Witko or 'Crazy Horse'. (chapter 3).

River Crossings in Australia and America

Although many travelling mobs were plagued by lack of water, particularly in Australia, but also in the United States of America, when water was encountered it had to be crossed.

In the days before Nat Buchanan pioneered the Murranji Track, Top End drovers using the old Gulf or Coast Track had many a scrape fording tidal rivers populated by sharks and large estuarine crocodiles which would have regarded a mob of swimming cattle as an unexpected smorgasbord. Smashes also occurred when thirsty stock watered too low down in tidal waterways and died from drinking sea water. Early drovers had to hold thirsty cattle off salt water, no easy task, until they got above the tide level to fresh water.

If cattle tried to turn back in mid stream, a mill could result which ended in a churning maelstrom that could cause deaths to men and stock. Cattle and horses will both try to climb over other animals or men in the water, and desperate situations can arise when river crossings go wrong.

To guard against a mill, most drovers would cross a small cut of cattle first and hold them on the far bank as coachers to draw the main mob across. A man on a good swimming horse always took the lead to give the cattle something to follow. Experienced trail drivers in America were known to take lead steers with them trip after trip because of their value at river crossings and also because they were accustomed to following the lead or 'point' riders as they were known in America during each day's travel.

Charlie Goodnight had a steer known as 'Old Blue' who made many trips up the cattle trails and always returned to the home ranch with the horses and cowboys after each drive.[18]

Dudley Snyder who made the two year drive to San Francisco, drove Texas cattle north to supply the confederacy during the Civil War. He had a yoke of work oxen that were wonderful swimmers which he would take out of the wagon and put in the lead of the herd at river crossings.[19] Needless to say they were too valuable to sell at trail's end and always pulled his supply wagon back to Texas.

The Greatest Stock Swim

The most amazing river crossing feat ever, happened during Dudley Snyder's era when Texas steers were being driven to supply the armies of the South. In October 1862 a raw-boned young Texan named W.D.H.Saunders left Goliad, Texas with 800 Longhorn steers and six companions, bound for New Orleans at the mouth of the Mississippi River.[20] The men of Texas were nearly all fighting for the Confederacy and the supply of beef to the troops was in the hands of boys as yet too young to wear the grey uniform.

Saunders was seventeen years old and as he filled the role of trail boss we can assume he was one of the older members of the party. History only records two other names of these almost beardless boys who drove the herd north east towards New Orleans. Monroe Choate and Jim Borrum. Part way the boys threw in with a small herd of 300 in charge of two more lads named Crump and Flemming, and the youngsters now numbering nine, pushed on towards the Mississippi delta.

The teenage trail crew swam the Colorado, Brazos, Trinity and Neches Rivers as they made their way across Texas. They swam the Sabine emerging on the eastern side in Louisana and pushed on towards New Orleans. Devastating news hit the youthful party part way across Louisiana, when they met a troop of Confederate cavalry. When the Southern officer found their destination, he promptly arrested the boys and impounded the herd. Unknown to Saunders and company, New Orleans had fallen to Union forces, but when the lads claimed with fire in their eyes that they would rather be shot than let so much as one steer go to the hated northerners, the officer released them and let them continue with a warning to stay well clear of their previous destination.

The boys considered their options and the decision was made to swing north and take the cattle round in a loop across Eastern Louisiana and Mississippi to Mobile, Alabama. They took the new direction and eventually the cattle spelled on flood plain grass and the young cowboys stood appalled at the task they had undertaken on the west bank of the 'mighty Mississip'. There was no way the cattle could be ferried and the only options were quit or swim.

An account of the incident records W.D.H.Saunders leaning on his saddle horn as he stared across the lower Mississippi, 1.6 kilometres (1 mile) wide, 12.2 metres (forty feet) deep, brown and sullen with tree branches and other debris sliding past in the current.

"Well boys. We got her to do and we can't let them damn Yankees get one over on us" he stated flatly, but we must wonder if there was a fearful tremor in the voice of young W.D.

The lads moved the cattle a few hundred metres back from the bank, took their ropes down from their saddles to use as whips, and with Saunders in the lead, 1,100 Texas longhorn steers, nine teenage boys and the wiry little mustangs they rode, charged the lower Mississippi River.

With whoops, yells and flogging rope ends, they ran the herd at the bank hard on the heels of Saunders as he spurred his horse into the river. The lead steers propped on the edge but the weight of the cattle behind forced them off the bank and all but 100 took to the mile long swim. When they hit the main current a small floating tree nearly turned the leaders back but Saunders dragged his horse's head round and came in on the upstream side of the lead steer with his heart in his mouth. The leaders followed him and once again they struck out for the east bank.

This swim of Saunders and his companions must have been one of the longest miles in history. The cattle and horses were swimming so low their noses were barely out of the water in the last hundred yards, but they made it. The triumphant cowboys alongside and behind the herd had to get up to the leaders and move them out of the shallows to let the others land they were so exhausted and locals on the other bank who sold the hundred which wouldn't cross on the Texan's behalf, said the sight was like a herd of upside down rocking chairs with only noses and horns visible above the water.

W.D.H.Saunders and his men drove the cattle on across Lower Louisiana and after another run-in with suspicious Confederate troopers, they got them to Woodville, Mississippi where they sold the original herd leaving Crump and Flemming who elected to drive theirs on to Mobile. Their story must rank as one of the most courageous droving feats ever performed.

A Bad Dry Trip

In 1866 when Texas was abuzz with news of Kansas cattle prices and every man and his dog was putting a herd together, Charlie Goodnight had another plan. We have dealt with Charles Goodnight's history as a cattle baron in chapter 5 volume 1, but before he began accumulating land and cattle, the ex-Illinois farm boy who migrated to the Brazos country in Texas as a lad, was one hell of a cowboy.

General Carleton had fought a long campaign against the Apache Indians bringing many into reservation life. Mountain man, scout and army officer Colonel Kit Carson had rounded up the Navajos in the Canyon de Chelly. This made for a large number of Indians at the Bosque Redondo Reservation near Fort Sumner, New Mexico dependent on the Government for food.

Beef was expensive in New Mexico that territory being mainly sheep country then, and Charlie Goodnight decided to tap that market instead of joining the rush to Kansas. To drive west was suicidal as the Comanche Nation lay right in the way. Charlie was well acquainted with this tribe from his days as a Texas Ranger. Like every south westerner, he had a healthy respect for these savage horsemen who were recognised as the finest mounted warriors of all the Plains Indians.

An old, Butterfield stagecoach road known as the Horsehead Route, route No.4 in the Principal Cattle Trails of the USA map following, lay to the south-west. It was in a region so desolate the short-lived stage line carted water to their change stations. The Comanches had better things to do than bother going there much and Goodnight determined he would drive across it. The biggest problem was a one hundred and fifty-four kilometres dry stretch (ninety six miles) between the middle Concho River and the Horsehead crossing of the Pecos. Could cattle do it?

Charlie and his partner Oliver Loving shared a great faith in the endurance of Texas longhorns. Goodnight had worked among them from boyhood and Oliver Loving was America's most experienced trail driver at the time. The thirty year old Charlie had sought Loving's advice on the matter and while warning of the dangers involved, Loving, a man in his fifties, was so intrigued by the daring plan, he asked to be involved.

This proposed march is reputed to be the birth place of that American ranching institution, the chuckwagon. Charlie designed the vehicle, and had it built from the tough wood of the Bois d'arc or Osage Orange and equipped it with several water barrels. Ten yoke of oxen were procured to pull the prototype and nineteen well armed, trail hands, all veteran Confederate soldiers (in case of Indian trouble), were employed.

The herd of 2,000 mixed cattle were put on the road in June 1866 and as events would soon show, mixing cows and calves with big mature beef steers (what we would call bullocks in Australia) was a bad mistake and one Goodnight would never repeat. Why two experienced cattlemen like Goodnight and Loving even considered taking cows and calves over a track like that is unexplainable, but they did, and they paid for their lack of foresight.

At first they tried carrying exhausted calves in the chuck wagon but it soon became obvious that the only way out was to kill the poor babies. This decided Charlie and Oliver to never again take mixed cattle this way if they managed to get this lot through.

The herd reached the middle Concho where they were rested and watered for the one hundred and fifty four kilometres dry trek they faced. Once under way, the cattle plodded all day across the high plains. Bitter alkali dust boiled in clouds forcing the cowboys to mask their mouths and noses with wet bandannas and gumming the animals' eyes, nostrils and mouths.

They bedded the animals the first night and got under way at daylight. Another day passed amid clouds of dust and that night the cattle refused to settle. They milled and tried to walk off the bed-ground until Goodnight and Loving had a conference in the early hours. They decided if the cattle were going to walk they might as well walk towards the Pecos as in circles, so the drive kept going.

At one stage the big steers in the lead got so far ahead Charlie put a bell on one of the point riders' horses. When the tail, or drag, as it is known in America, could no longer hear the bell, they sent a rider with instructions to hold the lead until the rest caught up.

By the third night, cattle were beginning to die and a dreadful low moaning came from the suffering herd. The only chance was to keep going and allow the laggards to drop out.

On the fourth morning they came through a pass in low hills called Castle Gap and sighted the Pecos nineteen kilometres away. When they got closer to the river, the cattle stampeded, when they smelt the water. When they reached the Pecos the leaders were pushed right across by the weight of the cattle behind and had to circle back to drink.

Three hundred head had died on that terrible stretch of desolation not counting the calves that had to be killed. A further one hundred were trampled and drowned in the charge into the river but Goodnight and Loving had learned from it and future droves of big steers or dry cows traversed the Horsehead route with minimal losses. Never again did they make the mistake of trying to camp the cattle either.

Principal Cattle Trails of the Unites States of America

Chisholm Trail	No. 15	Nelson Story/Bozeman Trail	No. 11
California/Arizona Trail	3	Oliver Loving Trail	7
Dodge City Trail	10	Oregon Trail	8
West to California (from Ogalala)	1	Phil Kearney Trail	13
West to California (Southern Route)	2	Santa Fe Trail	6
Goodnight Loving Trail	5	Sedalia Route	12
Horsehead Route	4	Shreveport Trail	14
		Western Trail	9

In spite of losses the beef steers fetched 8c a pound on the hoof at Fort Sumner – an unheard of price for Texas cattle, and Oliver Loving took the cows on north establishing the Goodnight Loving Trail, route No.5 in the Principal Cattle Trails of the USA map, to near Denver, Colorado where he sold them to the redoubtable John Wesly Iliff as breeding stock.

Goodnight packed $12,000 in gold coin on a mule and headed straight back to Texas to gather another herd. History does not record if Charles Goodnight was a profane man but it would be hard to believe that a few choice words did not fly when the mule bolted one night and almost lost the proceeds of the drive. Charlie managed to round up his errant pack animal, however, and of the second herd he brought across the Horsehead, losses totalled a mere five head. Charlie Goodnight was like Sid Kidman. He learned from his mistakes.

Women of the Stock Routes

The only reference I have been able to find to women on cattle drives in America is of a Mrs. Amanda Burks who accompanied her husband W.F.Burks on the long haul with a trail herd from Nueces County, Texas to Newton, Kansas in 1871 [21]. Mrs. Burks helped fight off Indians and rustlers as well as being exposed to other dangers of the trail such as thunderstorms, a prairie fire, flooded rivers and a stampede. Amanda is said to have driven a buggy on her long march north.

Australian stock routes saw a few women drovers over the years. One was Mrs. Richardson. In 1892 Nat Buchanan took cattle from Hugh and Wattie Gordon's *Elvire* block to establish a stock route to southern West Australian markets.

That drive established the Buchanan Track from the Ord River to Beringarra on the Murchison and the next year, Nat put 900 steers (young bullocks) from the *Elvire* block on the new route in charge of drover W.Richardson. Richardson was accompanied by his wife, a Malay woman, and three or four native stockmen.[22] In the early stages of the trip Richardson was hurt when he parted company with his horse and Mrs. Richardson assumed the duties of drover in charge.

Dressed in men's clothing and riding astride, the gutsy lady bossed the drove while her injured husband rode in the wagonette. Mrs. Richardson made a full delivery at Cue on the Murchison, six months after the drive started and in much the same manner as with Willie McDonald's heroic Chinese cook, the chroniclers of European history did not even bother to record her name.

During the years of the Second World War, Mrs. May Steele and her husband Jack drove bullocks from the Northern Territory to Queensland.[23] On one trip they lost their men and had to battle on with the cattle alone working all day and with long night watches.

Bush people have always risen to the occasion in times of need and the Steeles did not see anything special about their epic trips. The bullocks had to go through and that was all there was to it.

A road mob of mixed Willaroo, Manbullo *and* Delamere *'stores' string into the Ranken bore across the Northern Territory's vast Barkly Tableland in the 1950s. Only three men were with this mob of 1,250 bullocks. Bruce Simpson who took the lead and the photo, Johnny Shonrock on the tail and Jeff Simpson with the horses.*
Photo courtesy Jeff Simpson collection

The best known of Australia's lady drovers is Edna Zigenbine who delivered her first mob after her father Harry became ill with a mob of Quilty bullocks in the middle of the Murranji track. Some fanciful accounts claim Edna was only fourteen at the time, but she was twenty-three and already a veteran of the stock routes. She was the horse tailer on this trip and had certainly been on the road with stock many times since her early teens.

A point worth mentioning here is that cattle drives in America record the wrangler who looked after the horses, as a youth who occupied a lowly role in the hierarchy of the drive. Australian droving tradition lists the horse tailer as having a highly responsible job second only in seniority to the drover in charge, or boss drover as they have become known.

The bulk of American cattle trails were in well grassed and watered country, at least by Australian standards, so the wrangler would only have to drift his charges, the cavvy or remuda as they were known, along far enough from the cattle to get grass. The supplies and bedding were carried in wagons and the wrangler's main responsibility was to have horses ready when cowboys wanted to change mounts. American cattle drives also employed another lad called a 'nighthawk' who minded the horses at night and helped the cook travel with the wagons and set up camp.

In Australia, most droving plants used packhorses on long trips with store bullocks, so they could venture into rough country unsuitable for wheeled vehicles, in search of grass. The horse tailer's job was to catch the riding and pack horses and load the packs each morning, find grass and water for the horse plant and assist the cook setting up the camp. When grass was scarce in drought times a horse tailer would often take the horse plant miles to find a patch of feed and camp with them returning to the cattle in the early hours. The success or failure of a droving trip depended on the horses and horse tailing was a vital job. Boss drovers considered the welfare of the cattle, the horses and the men always in that order.

The fact that Edna Zigenbine was horse tailing on the notorious Murranji Track, when forced to take over the mob, means she was no novice and in spite of walkouts by her men, an outbreak of pleuropneumonia in the cattle as well as the usual hazards of droving, she made her delivery in Dajarra, Queensland and went on to a successful career as a boss drover [24].

In early times many drovers preferred Aboriginal women to their male counterparts for stock work. Apart from the more obvious attractions, they were believed to be more industrious and better with horses and cattle. Many nameless women have ridden Australia's cattle tracks and female names given to creeks, yards, springs and soaks are the only memories left of those dusky overlanders.

The End of an Era

Big mobs walked out of Australia's back country from Joseph Hawdon's time in the 1830s until the mid 1960s when road transport took over. Around 1960 drought closed the major stock routes. Road trains had been operating since the late 1940s when *Marion Downs* in Queensland's Channel Country began experimenting with stock transporters. With droving impossible, the trucks took over and were the main method of moving stock from then on. This set of circumstances brought the golden years of droving to a close by the mid 1960s, but drought or not, the shift to road trains was inevitable.

In recent years droving has become common again as big cattle companies find it is cost effective to walk stock between their properties. This has the added advantage of being a very good means of quietening and educating station bred cattle.

As long as Australia has a pastoral industry it will be stalked by the spectre of drought. Often the only option for cattle and sheepmen whose country is eaten out, is the 'long paddock' and particularly in dry years, it is always possible to see mobs 'on the road' in Australia.

DUFFERS & RUSTLERS

In the north when bearded bushmen yarning with their pipes aglow,
Tell the tales of roaring Roma, Roma sixty years ago;
Time has halted in their stories and the cattle bandits ride
Westward to the Maranoa in the wild years that have died.
Never in the Queensland story was a bolder epic born
Or an era passed as reckless in the zone of hoof and horn,
When on dim tracks rarely travelled, camping with the rising suns,
Stolen steers through starlit stages found their way to distant runs.

Lex McLennan

Duffing, Horse Theft and Poddy Dodging

Song of the Sheep Stealer

Australia's unofficial national anthem celebrates a small time stock thief, on the surface anyway. However, dig into the events behind Banjo Paterson's much loved melody of the swagman and jumbuck and a story of rebellion and civil unrest lurks beneath the seemingly frivolous words of *'Waltzing Matilda'*.

The sheep allegedly stolen in the song was in reality 140 lambs burnt to death when militant shearers fired the *Dagworth* woolshed near Kynuna, Queensland on 2nd September 1894 in the final stages of the 1890s shearers' strikes. (chapter 4)

The Swagman, one Samuel 'Frenchy' Hoffmeister, committed suicide by a billabong about eight kilometres from Kynuna – not the Combo waterhole, a man-made weir, that has been given credit as the site. Hoffmeister, who suffered from mental illness, took his own life, probably due to remorse over the dead lambs which had no part in the plan. The strikers had not meant to harm the lambs when they attacked the fortified woolshed and it was regretted by the strikers as much as the squatters, station hands and police who defended it[1].

When Paterson penned the words to *'Waltzing Matilda'*, it was not long after his friend and fellow poet, Henry Lawson narrowly escaped treason charges after the release of his controversial ballad *'Freedom on the Wallaby'*. Banjo hid the true meaning of his song in the seemingly light hearted lyrics, but *'Waltzing Matilda'* is really a freedom song commemorating a working class struggle.

If a swagman who supposedly pinched one sheep (and how he managed to stow a live, western wether in a one-man tuckerbag is beyond me) became the theme of one of the world's most loved songs, what could the Banjo have done had he chronicled the exploits of some of the more fair dinkum stock stealers?

Harry Redford

Australia's most notorious cattle duffer was Big Harry Redford or Readford, who became the partial inspiration for the fictitious character of 'Captain Starlight' in Rolf Boldrewood's classic novel *Robbery Under Arms*. 'Starlight' was the ultimate gentleman bushranger: courteous to women, honourable to gentlemen, a dashing scoundrel in the mould of Dick Turpin and Robin Hood. Harry Redford's droving feat was adapted to 'Starlight' by author Boldrewood.

Redford's story has been told in almost everything written about history of the Outback, so I will only deal with it briefly on the off chance some reader has not heard the story.

Henry Arthur Redford was born at Mudgee, New South Wales on 12th December 1841[2]. His family were noted landowners who were later scandalised by their wayward boy's exploits in long-horned larceny. Redford was described as a big, easy going man, well read and educated and an exceptional cattle and horseman.

In 1870 Harry and a few accomplices stole 1,000 bullocks and a soon to be very controversial white bull from *Mt.Cornish* Cattle Station, an outstation of *Bowen Downs* established by William Landsborough and Nat Buchanan after their exploring trip through central Queensland in 1860. Harry and his mates drove the cattle south west with the idea of selling them in Adelaide. As an epic of droving, it has few peers. They sold the bull to a station owner en route and finally disposed of the mob at Lake Frome in South Australia.

With the white bull as evidence, Harry was arrested two years later and tried in Roma, Queensland on 11th February 1873. It was apparently a bit hard to find 'twelve good men and true' in Roma in the 1870s or at least twelve who sympathised with large scale property owners.

Harry was acquitted in the face of overwhelming evidence of his guilt and scathing comments from the officiating Judge Blakeney. Roma lost its court privileges for two years as a result. The verdict may not have come about totally as a result of dislike for big companies and admiration for Redford's stockmanship. It was rumoured around stock camps throughout Queensland and the Territory at the time that Harry spent every penny he owned 'squaring' the jury[3].

Harry was carried shoulder high from the court and the general feeling of the community appeared to be that they would not see a drover and cattleman of Harry Redford's class rot in jail. Harry did finish up doing some time later on due to an inability to keep his hands off other people's horses.

Redford eventually drove breeding stock into the Northern Territory to establish *Brunette Downs* on the Barkly Tableland. He may or may not have gone straight after that, but if any of his larcenous instincts remained, Harry would have been in like company in the early days of the Territory. Many of the men out there in those times were convict sons of convict parents. Incidents of entire stock camps taking to the bush at the appearance of a police patrol were not uncommon in those wild times. Whether Harry kept his nose clean or not is immaterial. He certainly wasn't caught.

Australia's most famous cattle duffer achieved immense respect as a bushman, horseman and cattleman and was tragically drowned in Corella Creek in his later years while trying to free a horse whose back feet had become trapped in its hobbles in the water. He is buried near Corella Creek on the Barkley and his grave has been restored as a fitting monument to a man who, cattle duffer or not, was one of the greatest stockmen Australia has ever seen.

Kimberly Cattle Duffer Jim Campbell

The history of Northern Australia has recorded Jim Campbell the cattle duffer, as practically everything from a Robin Hood of the north to a low and scabrous cur. Renowned drover and station manager, Dennis 'Jim' Ronan, once flew in the face of established bush tradition by refusing Campbell the hospitality of his camp during Ronan's time as manager of *Victoria River Downs*. It was an unwritten law of the Outback that all travellers be offered food and drink, and Ronan solved the problem of his conscience and reputation by walking out of camp and refusing to eat himself [4].

It is certain that Campbell fell from grace with his fellow poddy dodgers who lived by their own code of brand-altering artistry, and became no more than a blatant stock thief. On the other hand he was a superlative horseman, cattleman and bushman who learned the duffer's trade in a time when getting a kick along at the expense of the big properties was an accepted career move amongst working class battlers.

Jim Campbell was born at Tenterfield on the New England Tablelands of New South Wales, as Ernest John Muir in 1871. He was one of eight children in the family of John and Johanna Muir who moved around the northern New England region, most of the children being born at Tenterfield but some at Warialda and Inverell. Ernest was more commonly known as 'Sonny' and legend has him leaving home in a hurry a short distance ahead of a local policeman with some rather awkward questions regarding ownership of stock. Sonny was said to have learned his advanced horse skills at the hands of a masterful horseman named Jim Cummings while running brumbies in the McDonnell Ranges near Alice Springs [5]. He first appeared in the Kimberley region of northern Australia in 1894.

Sonny Muir passed through *Newcastle Waters* in the company of brothers Ben and Tom Martin. The trio had a splendid mob of thoroughbred horses with them and it is probably unlikely any of the animals were accompanied by a receipt. The horses had been bred on stations in central Australia involved in the Indian Remount trade where some of the country's best horses were bred. Brands noticed at the time were from several well known properties including Sir Thomas Elder's *Blanchewater* and *Owen Springs* Stations [6].

The name Muir had been discarded by the time the tall, sandy haired, young bushman who sported a magnificent flowing moustache arrived in the north. The alias he gave, Jim Campbell, was the name by which he achieved notoriety as a cattle thief. His real surname was known in the top end where his middle name was believed to be Clare. A search of records in the New England region proved this was incorrect and Jim Campbell had begun life as Ernest John Muir.

Campbell, in company with a man called Mick Flemming, took up a block of country known as *Illawarra* adjoining the huge *Victoria River Downs* Station. The two enterprising individuals began stocking their station with unbranded V.R.D. cattle that were soon decorated with Jim Campbell's brand, the diamond eighty-eight. If a few branded stock were driven along with the cleanskins, Jim's artistic talents with a piece of heated metal soon created a new identity for the relocated bovines.

Victoria River Downs was the biggest cattle property in the world from the time of its establishment in September 1883 until its area was reduced by resumption in the early 20th century. Its 32,000 square kilometres proved a thorn in the side of efficient management for many years. There were huge numbers of unbranded cattle roaming its sprawling reaches and this acted like a magnet to men of Jim Campbell's ilk. Despite the fact his mustering teams were unable to gather much more than half the cattle on the station, Manager Dick Townshend took a dim view when he discovered tracks of driven cattle and shod horses leading directly to *Illawarra* and he served a summons on Jim and Mick.

Buffalo shooter Ben Martin, who had come into the Top End with Jim, offered to buy a third share in *Illawarra* for £500. and swear the place was his and that the accused pair had been acting under orders. A sale deed dated six months previously was drawn up and the three partners rode into Timber Creek to attend the trial which coincided with a race meeting there.

As at most bush race meetings, the grog flowed freely in the lead up to the event and by the time the case was heard at Timber Creek Police Station, public opinion was squarely on the side of the two accused battlers. Almost everyone present agreed it was a pretty 'rum do' when a bloke got charged for putting his iron on a few cleanskins that the station hadn't been able to catch anyway. If the men who mustered them were game enough to risk a spear through the ribs from the wild blacks in regions the *V.R.D.* stock camps avoided, then they damn well deserved the cattle.

One of the Kimberleys legendary pioneers, Captain Jack Bradshaw, was the Magistrate and after Townshend and his stockman Jack Frayne gave evidence of following tracks to *Illawarra* and finding Campbell branding their cattle, he advised all parties involved to let the matter drop. A police constable present made a considerable fuss over Bradshaw's rather cavalier interpretation of the law and during the loud and profane squabble that ensued, Townshend's stockman Jack Frayne took his boss aside and told him if the case was sent to Darwin he would refuse to testify.

Townshend retaliated by threatening to refuse Jack further employment. Frayne grinned and notified the manager he intended to quit anyway.

In the next few months, Jack Frayne took up new country on the Western Australian border and the nucleus of his breeding herd was one hundred *Illawarra* heifers presented to him by Jim, Ben and Mick for services rendered. It only added to the joke that the heifers had started life in the herds cared for by Jack's former boss.

With defeat staring him in the face, the *Victoria River Downs* manager reluctantly withdrew his charges and everyone present at Timber Creek settled down to the serious business of celebrating the outcome and the race meeting, in true northern style.

Jim Campbell was far too cocky and he couldn't resist rubbing a bit more salt into Dick Townshend's wounds. He had his eye on a bit more country that lay against the big run and he wrote to the Lessee who was about to lose the place due to failure to stock his grazing lease. Jim offered to put cattle on the place to comply with the lease terms in return for the use of the 1,000 square kilometre block. This was a fairly easy promise to make as the place was already over-run with *V.R.D.* cattle. The Lessee, W.H.Kirby of Collarenebri, New South Wales who had practically forgotten he had taken up a northern run, agreed to transfer the lease. Jim moved onto his new place naming it *Retreat*. He left the running of *Illawarra* to Mick Flemming and Ben Martin and set to work harvesting unbranded cattle.

In constant danger of being knocked out of their saddles by a shovel nosed spear, Jim with Billy Linklater who was working for him on shares, and a few Aboriginal stockriders began mustering the cleanskins that swarmed on *Retreat*.

He sold his interest in *Illawarra* to his partners and when his share of the cattle was sold, Jim Campbell had more money than he'd ever had in his life before. A holiday was called for. Jim took ship for the south where elegantly clad in the latest fashion, the Territory poddy dodger lived it up in Sydney's best hotels before making the classic bushman's pilgrimage to The Melbourne Cup.

The Flemington bookies were financially unkind to Jim but several other 'Northern Terrors' were down for The Cup that year and one was more than happy to cover Campbell's losses and lend him enough money to get home. His benefactor laughed off the promise of cash and told Jim he would be happy to take the debt in cattle.

Jim Campbell had his own code of ethics and the racing debt worried him. While he had no compunctions about stealing from the big concerns, a debt to a mate was a sacred trust and Jim wanted to square up quick. It was around this time Campbell began his slide downhill from a popular larrikin who played the game according to the unwritten rules of the duffing fraternity, to an out and out stock thief shunned by his former friends.

To save himself the trouble of mustering clean skins, Jim registered a new brand 919 that fitted squarely over the top of the GIO of *Victoria River Downs*.

Campbell was now blatantly running off branded V.R.D. cattle and crudely blotching their brands[7]. Billy Linklater pulled out of the operation, in his words due to a difference of opinion over what constituted a cleanskin, and soon the only white man in the north who wanted anything to do with Jim Campbell was a runaway sailor named 'Dutchy' Behning.

Behning was known as a pretty low sort and his presence at *Retreat* was obviously to get his hands on part of the easy pickings Jim was driving in from V.R.D. Eventually he and Campbell had a flaming row. Jim king-hit the ex-sailor and 'Dutchy' rode to the big run to inform on his former partner[8]. Jim soon got the news that 'paper' (an arrest warrant) was out for him and giving up his dreams of pastoral wealth at the big run's expense, he left for the coastal country east of Darwin.

Police arrived at the abandoned *Retreat* where they found an Aboriginal woman who claimed she could lead them directly to their quarry. With the lady in the lead, the party headed north west to be told days later on a rocky hill near the mouth of the Victoria River "That fella gone away now". Jim may have been getting short of friends, but he obviously had one left.

Campbell settled into a kind of beachcomber existence along the north coast of Arnhem Land. He obtained shares in a lugger with a seaman named Charlie Williams, collected trepang, for sale, and shot buffalo for their hides. The police became aware of his presence there but knowing where he was and catching the wily outlaw were two different things.

Jim had a good plant of horses as well as access to a seagoing vessel and the natives always let him know of strangers in the region. If the police came by sea he would be in the ranges. When the patrols came overland Campbell would put to sea and eventually the forces of law and order gave up. They reasoned that if they could not catch Jim Campbell at least they had him yarded out where he couldn't get up to any mischief with other people's cattle.

Campbell had a reputation for being pretty hard on Aborigines who worked for him and it was this nasty trait in his character that was to finally bring him undone.

Coastal Trader Alf Brown who had transported trepang and buffalo skins to Darwin on Campbell's behalf, warned him repeatedly that his treatment of Aborigines would end in trouble and when Jim dished out a hiding to one of his blacks and knocked him into a vat of scalding water used for cooking trepang, the tribesmen took matters into their own hands.

Campbell pushed off from his lugger one night in a dugout canoe near Guion Point on the coast of the Arafura Sea. He and his native helpers along with a half-caste named Tommy Carpenter worked their way up a stream on a mudflat surrounded by mangroves and began collecting bêche-de-mer. Sometime around midnight, a chilling war cry rang out over the night noises and slap of wavelets against the canoe as tribal avengers loomed out of the darkness into the flickering light of Jim's paper-bark torch. Campbell lunged for the dugout where his rifle lay. The white-ochre-daubed assassins hurled their spears, one taking Jim in the throat and eight more finding their mark in his body.

The outlawed white man lay writhing in the mud as warriors mad with blood lust, seized the paddles from the canoe and shattered his skull ending the life of the north's most notorious cattle duffer.

Tommy Carpenter and the other natives being unarmed, could only stand helplessly and when the tribesmen fled, they returned to the lugger. Campbell's partner, Charlie Williams, came back in the morning with all the crew and they cleaned Jim up, dressed him in clean clothes and buried him sewn in a spare sail.

They sailed for Darwin to report the murder and trooper J.R.Johns and another named Cameron, sailed on the *'Maggie'* to investigate. By questioning the coastal tribes they learned who the nine murderers were and when they reached the creek near Guion Point, Campbell's body was disinterred for formal identification and reburied in one of Australia's loneliest graves near the mouth of the King River.

In Darwin Supreme Court during September of 1913, Warditt, Angoodyea, Terrindillie, Dariilba and Temerebee were sentenced to hang for the murder of Ernest Clare Campbell, trepanger, and their four accomplices were released.

A strange twist to the story is that Jim Campbell could have returned to white society as a free man because unbeknown to him, the arrest warrant had been withdrawn only a few days before his demise.

The Keane Tyson Brand Affair

When registering a brand in earlier times, a few enterprising individuals managed to gain the right to use letters and symbols that were ridiculously close to those in use by their neighbours. This situation was later rectified when all applications were required to have approval from neighbouring properties.

In much the same way as Jim Campbell managed to obtain his **919** brand that fitted fairly neatly over the G I O of *Victoria River Downs* and was eventually responsible for the big run changing their brand to the bull's head symbol, millionaire grazier James Tyson once found himself in a similar situation.

Tyson's huge *Tinnenburra* holdings sprawled on either side of the Queensland/New South Wales border in the area around Cunnamulla and in keeping with his status as the number one grazier in the country, Tyson's brand was **T Y 1**. A neighbour whose property adjoined Cunnamulla and like the town was surrounded by Tyson's holdings, went by the name of Timothy Keane. Tim Keane was a noted scrub rider and brumby runner who was also a mail contractor for the run from

Cunnamulla to Barringun just over the N.S.W., border when he first came to the Warrego region [9].

Keane went in for carrying later and took up a selection three miles south of Cunnamulla where he bred good horses and registered his TK7 brand which became a thorn in the side of James Tyson when he put the huge *Tinnenburra* run together. (chapter 5 volume 1).

By the simple act of adding an extra two strokes to the *Tinnenburra* brand, Tim could convert ownership of stock on all of his boundaries to his *Cuttaburra Branch* holding. Keane knew it and Tyson knew he knew it.

Timothy Keane had no wish to sell his holding and refused all offers, so Tyson reasoned if he could buy the brand itself, it would remove the temptation for Keane to enrich himself at Tyson's expense. He sought out his neighbour and made an extremely generous offer for the times, of £1,000 for the purchase of Keane's registered brand.

It would be fair to assume that a certain amount of enrichment had already taken place because Keane laughed in the big grazier's face and said "Ah don't be bloody silly Tyson. This brand's worth a thousand pounds a year to me."

Henry Beecham and the Thoroughbred Brumbies.

Henry Beecham was a specialist horse thief who drifted onto Queensland's Darling Downs in the early days of white settlement. Adept at faking brands and hiding stolen stock in out-of-the-way places until owners gave up hope of their recovery, he did quite well at his chosen profession until a trooper named Armstrong caught him red-handed in possession of a stolen horse.

Three years in Brisbane jail failed to cure his partiality for other people's steeds but it certainly gave him time to work out what Henry considered to be his master plan.

Soon after Beecham was released, well-bred mares began to disappear all over the eastern Darling Downs in large enough numbers to fuel rumours that an organised gang of horse duffers was marauding in the region. The 'gang' was one man, Henry Beecham, and instead of running the considerable risk of trying to sell or travel animals of such obvious quality, he was secreting them in a hidden valley in the Main Range near Pilton east of Toowoomba.

All the normal avenues were investigated by police with particular attention to race meetings and sales throughout the region that would become south east Queensland and northern New South Wales. There was no trace of the missing mares because they were secure in Beecham's mountain stronghold.

The next link in the chain of thefts came when *Clifton* Station reported the loss of '*Clandestine*' their prized thoroughbred stallion which had disappeared from his stable in the dead of night during a violent thunderstorm. Rain had obscured all tracks and the valuable animal appeared to have dropped off the face of the earth. The thefts stopped and gradually horse breeders across the Downs began to relax their vigilance.

Henry Beecham left the district, his unofficial stud established in the hidden valley near Pilton, and went to Brisbane to wait out the three years until his scheme bore fruit in young unbranded thoroughbreds ready for breaking in. He obtained work on a farm just outside town but Beecham, like most criminals, was unable to keep his hands off other people's property. He foolishly stole another horse. The police moved quickly and Henry once again became a guest of the Crown for a considerably longer stay this time.

Ten years passed before stockmen roaming the ranges reported a mob of magnificent horses in a pocket in the hills ruled by an old chestnut stallion with saddle marks and a well defined brand. James McGregor, head stockman on *Clifton* Station put two and two together and when taken to the hidden valley he recognised the missing *'Clandestine'* at a glance.

The thoroughbred horse faced no competition from wild stallions other than his own sons in those days as horses were scarce and well husbanded. It was later on in the last two decades of the 19th century, the brumby mobs that still exist in mountains adjoining the Darling Downs became established. *'Clandestine's'* wild progeny would have formed some of the foundation stock for the wild herds but when Beecham's stolen mob and their progeny were discovered, they were pure thoroughbreds untainted by outside blood.

McGregor arranged with the owner of the country to erect yards in the pocket and after some unsuccessful chases in the beginning, he and his six helpers eventually yarded the stallion and a few others by running them in relays and wearing the fiery wild horses down. *'Clandestine'* was returned to *Clifton* Station where he ended his days. For years after, some of the brumbies caught in the mountains around Pilton were of sufficient quality to gladden the heart of any horseman.

Galloping Jones, the Ultimate Larrikin

John Decey Jones, better known as 'Galloping' Jones, was a legend in North Queensland in the first half of the 20th century, but finding concrete evidence about him is like following a stony track up a lot of dry gullies. It seems fairly sure Jack Jones was a returned World War I soldier who should have been active years before in the heyday of the poddy dodgers.

He had a grand disdain for rules, lived by the personal code that he could do anything he bloody well liked and took a backward step for no man. Jones was a top stockman, a well above average rough rider who travelled for a time with Lance Skuthorpe's Wild Australia Show, a talented axeman and an extremely tough man to beat in a bare knuckle scrap.

He was of average height and weight but fit, athletic and extremely game. 'Galloping' was also possessed of a larrikin personality and many of the stunts he got involved in seem to have been motivated more by a desire to thumb his nose at society than any personal enrichment that may have resulted. He and his mates certainly did a bit of duffing and killed meat for themselves or for sale whenever it suited them, but according to stories told, the 'Galloper' never 'stole' horses.

He may not have been guilty of outright horse theft, but he certainly had no objection to borrowing horses without their owner's knowledge or permission. Neither did it trouble his conscience to sell those 'borrowed' horses when the opportunity presented itself, and owners who wished to reclaim animals they were unaware they had 'loaned' the 'Galloper', faced the problem of finding them first.

The story goes that when confronted in possession of a horse by its legal owner or police, Jones would grin disarmingly and willingly surrender the animal. Getting him to accompany the borrowed horse and its recoverers to face charges was a somewhat tougher proposition. This practice, commonly known as horse sweating, could only be dealt with by a charge of Illegally Using, and big stations, which numbered their horses in thousands, mostly adopted a lenient view knowing the animals in question would turn up somewhere sooner or later [10].

'Galloping' Jones didn't care if owners knew he had their animals or not and a story exists of how he once tackled a well bred outlaw that had thrown all the local crack riders. Jones mounted the horse in the station yards and after easily sitting the first few twisting bucks he yelled to open the gate. The horse plunged through the opening, coiling like a demon to the accompaniment of yells from the assembled watchers to 'Stick to him 'Galloping'!'

Jones tossed the remark back over his shoulder that he'd stick to him alright and stick to him he did, all the way to a hidden camp he knew in the ranges near Charters Towers.

One of the most common anecdotes told about 'Galloping' Jones is how he came by his nickname. Apparently Jack Jones and a few of his mates worked out a scheme to ring-in a fast horse at a bush race meeting. In those times, it was still legal to dock a horse's tail, and the animal in question was a 'cob, as docked tail horses were called. Jack and his cobbers manufactured a false tail and with a bit of boot polish applied to white legs and face markings, an excellent job of disguising the speedy beast was arranged.

The horses lined up for the start with Jack riding the ring-in. The flag fell and the field surged away. Jack was well placed as they came into the straight. The money was on with the bookies at very attractive odds and his mount answered hands and heels drawing away from the field as they approached the winning post.

Suddenly, Jack Jones heard a change in the pitch of the crowd's cheering as the excited babble began to change to a roar of anger and he glanced over his shoulder to see the artistically constructed, false tail parting company with his mount.

Jones made the decision then and there to have a look at a bit of new country. He galloped past the post and just kept galloping off the course, out of the grounds, and away earning the name that would stick to him all his life [11].

Always at odds with the police force, Galloping Jones was not adverse to throwing down the gauntlet to them. The Queensland coppers wore khaki uniforms in those days and cattle country towns throughout the north saw the 'Galloper' mount his horse and race up and down the main street roaring "Yard me you khaki bastards! Yard me!"

Jones was yarded on quite a few occasions and one of the few pieces of solid evidence I have been able to find is his mention in the Police Gazette. On 29th May 1925 he was committed for trial in the Cairns Circuit Court at Chillagoe on a charge of assault occasioning bodily harm to Herman Carl. The name of his victim suggests German ancestry, so perhaps the 'Galloper' was still fighting The Great War.

He must have been in exceptionally good form at Capella on 18th July 1927. This time, charges listed in the wake of one of his main street challenges to the law, included creating a disturbance, resisting arrest, obscene language and assaulting police. Jack was back in trouble in 1928 when he did a three month stretch in Townsville's Stewart's Creek Prison for disobeying a maintenance order for his wife and children. His description was circulated for the same offence in July 1930.

The description given in the Queensland Police Gazette 19th July 1930 was as follows:

John Decey Jones otherwise Galloping Jones
otherwise Alexander Brown, 36 years of age, 5 ft 6 ½ in.
high, dark complexion, brown hair, clean shaven,
brown eyes, regular nose, scar on left shoulder,
minus two joints little finger of right hand, bald on top
of head, a horsebreaker, native of NSW

The 'Galloper' was once shot for resisting arrest. This caused the shoulder scar mentioned in his police description. The cops had him in custody and were riding on either side of him holding a rein each while Jones sat in his saddle with his hands cuffed behind his back. The conversation between the policemen turned to racehorses and Galloping suddenly broke in with, "Watch this bastard race" and drove his spurs into his horse's sides. The reins were pulled out of the officers' hands and Jones went bush at a hell of a pace. It was a masterful display of riding as he swayed with the movements of the horse unassisted by reins or hands as it tore through the trees leaping over rocks and fallen logs.

One of the pursuing policemen snapped several shots at the 'Galloper' with his revolver and he must have been a fair enough marksman. A bullet took Jones in the shoulder and lifted him out of the saddle necessitating a few days in hospital before quite a few more in the lock-up [12].

'Galloping' Jones lived out his life according to his own rules and even as an old man living in the Eventide Home for the aged at Charters Towers, Queensland, he had a habit of going AWOL and becoming involved in brawls and drinking sessions.

He passed away in Charters Towers around 1960 but stories of his larrikin life live on particularly in North Queensland, his old stamping ground. Had he been born in an earlier time, Jack Jones may well have gone down in history as one of Australia's celebrated outlaws. 'Galloping' Jones left a legend behind that entitles him to be known as the Last of the Old time Poddy Dodgers.

Stewart's Creek Prison, Townsville in the early 1900s viewed from No. 3 Tower. Galloping Jones was a guest of the Crown here.

Courtesy Taylor Family Collection

Duffers of Today

Stock theft is by no means an old fashioned occupation. On both continents modern transport has made it possible for stolen animals to be moved long distances in very short times. Rustling and duffing rings turn up occasionally with property owners, truckdrivers, stock agents and inspectors all "in the game" of moving misappropriated animals by the semi-trailer load.

In April, 1989 three men were jailed for seven years over the theft of 6,000 head of cattle in Queensland's Gulf Country. This would have to be one of the biggest stock thefts of all times.

Rustlers in America

The Tombstone Cowboys

The most publicised gunfight in the history of the west and the world, occurred at the O.K. Corral in Tombstone, Arizona on 26th October 1881[13] when Billy Clanton, and Frank and Tom McLaury died under the guns of Wyatt, Morgan and Virgil Earp, and their friend John 'Doc' Holliday. Popular history recorded it as a classic confrontation between the forces of good and evil. A lesser known fact about the famous encounter is that it rang the death knell on one of America's most organised and notorious stock theft organisations.

In accurate accounts of the conflict, it can be seen the boundaries between good and evil were blurred to say the least. The Clanton McLaury faction were members of families who moved into the area that would become Cochise County after the first settlers had established cattle ranches well before the mineral finds of the late 1870s. Tombstone was one of the last wide-open places and it attracted a lot of frontier riff-raff as the spread of law and order made life difficult for outlaw factions across North America.

The Clantons ranched at Lewis Springs near Charleston, a small town about fourteen kilometres south west of the site of Tombstone. The McLaurys had settled in the Sulphur Springs Valley west of the spot where the boom town would spring into being following the discovery of silver by Ed. Schieffelin in 1877. Once the silver rush started the demand for meat became more than local breeders could supply and a rustling element evolved with the Clanton and McLaury ranches as its headquarters.

The head man of the Clantons carried the initials N.H. for Newman Haynes, but history has mostly recorded him as 'Old Man Clanton'. Vigilante forces had run him out of California during the gold rush era and his three sons Ike, Billy and Phin, were raised rough with no respect for any law but the gun. The McLaury brothers, Frank and Tom, were known for their prowess with firearms and the crowd that gathered at the two ranches was as choice a gang of cut-throats as the west had assembled.

Among the more notorious members of the Clanton/McLaury faction were Johnny Ringo, Pony Deal and Curly Bill Brocius. They were a mixture of cowboys who drove herds to the new rush and stayed, remnant fighters from New Mexico's Lincoln County War that thrust Billy the Kid into prominence, escaped criminals and border scum. All told around 300 active desperados and hangers-on formed a loose alliance who called themselves 'The Cowboys' and lived by stealing cattle and horses in Mexico for sale to the miners. This was the faction history would record as the 'Bad Guys'. It was a fair enough assessment. The Clanton and McLaury boys and their followers were a nest of uncouth rowdies who regarded Tombstone as theirs by right of prior occupation, and whooped it up in the new town with a vengeance.

The Earps who would become known as the 'Good Guys' were no angels either. It's true Wyatt and Morgan had worked at law enforcement before their arrival in Tombstone, but Wyatt had been charged with horse stealing near Fort Gibson, Oklahoma, and the family's main occupation seems to have been saloon keeping and gambling. The Earps were also known to dabble in prostitution with Wyatt's elder brother Jim and his wife Bessie having kept a brothel in Wichita, Kansas. They were sometimes called the 'Fighting Pimps'. Their friend 'Doc' Holliday was a consumptive dentist who had come west for his health. He was an alcoholic known to be a 'mean drunk' with several killings to his name.

The gunfight at the O.K. Corral came about as the result of a personal feud between the two factions fuelled by political aspirations to run Tombstone, and corrupt officials on both sides. Tombstone put up with 'The Cowboys' at first. Their stock- stealing activities took place south of the border and they spent freely in town. The miners needed meat and very few North Americans in those days gave a hoot in hell what happened to the Mexicans or 'Greasers' who were the target of the rustlers.

Things got a little hot for raiders who swooped across the border into Sonora intent on pillage and quite prepared to shoot any vaqueros who got in their way when Old Man Clanton and six others lifted a herd of Mexican steers and pointed them north.

Clanton and his henchmen drove the cattle into Guadalupe Canyon that connected Sonora and Arizona, and bedded them for the night about 200 metres into the United States. Mexican troops were not supposed to cross the border, but after eight Mexicans had been killed by 'cowboy' raiding a few days previously, the party of Mexican regulars led by Captain Carrillo who pursued the rustlers, were not inclined to let 200 metres cheat them of their prey. As Clanton and company lay in their bedrolls the Mexicans opened up, killing all but two: Harry Ernshaw and William Byers.

The leadership of the Cowboys passed to Curly Bill Brocius and it was then that depredations began on ranches north of the border. The gang also branched into stage coach robbery; travellers were murdered for their money and goods, and some law officers in Tombstone were soon in the cowboys' employ. John Behan, Sheriff of Cochise County, was definitely in cahoots with the rustler element and he also had a personal reason to hate Wyatt Earp. Wyatt had superseded him in the affections of Josephine Sarah Marcus, Behan's common law wife.

Wyatt Earp and his brothers became embroiled in this volatile situation. The legendary gunfight at the O.K.Corral has been lauded in books and movies ever since, and was the opening battle in the 'Cowboys'/Earp feud that broke the Tombstone rustling ring.

Some weeks after the O.K.Corral, Virgil Earp was shot from ambush and wounded. Five months later another unknown assailant shot Morgan Earp dead through the window of a saloon where he was playing billiards. If Wyatt Earp had been prepared to let bygones be bygones before, which appears likely, things changed then, and he carried his blood feud to the 'Cowboys' with deadly efficiency.

Within two days he had tracked Frank Stilwell to Tucson and killed him believing Stilwell to be the man who killed his brother. The killing was in fact done by Pete Spense whose real name was Lark Ferguson. Spense was later arrested and sent to Yuma State Penitentiary for the crime. By Wyatt's own admission, he shot Stilwell in cold blood while he was begging for his life [14]. Another Earp brother, Warren, made a fast trip from California to join the hunt while Wyatt recruited gunmen for a posse that was technically legal but would soon overstep the limits which governed police forces.

Texas Jack Vermillion, a flamboyant, long haired frontiersman, was along as were Sherman McMasters and Jack 'Turkey Creek' Johnson who had been 'Cowboys' on the surface but were in reality plants for the Earps and the law and order faction in Tombstone who backed them. The sinister 'Doc' Holliday was also one of Wyatt's men and they were later joined by Charlie Smith. The fighting Earps and their quasi vigilantes scattered the rustler faction like chaff.

Ike and Phin Clanton fled to Mexico with Pony Deal, soon to meet their nemesis at the hands of Sheriff Commodore Perry Owens as described in chapter 2 volume 1. Curly Bill Brocius, the captain, must have felt honour bound to figuratively 'go down with his ship' and go down he did when he met Wyatt Earp in a face to face duel, both of them armed with shotguns. Curly Bill got the first shot away but quick does not mean accurate, his pellets only tearing the long coat Wyatt wore. Earp took no chances when he cut loose and Brocius died in a swarm of double O buckshot.

The Earps and their relentless fellow avengers swept through Cochise County like a bad wind. Gang members Florentino Cruz better known as Indian Charley, Johnny Ringo, Harry Head, Jim Crane and Billy Leonard died with their boots on in true western tradition and the remnants of the Tombstone 'Cowboys' scattered to the four winds.

Wyatt Earp, unlike many of his gun fighting contemporaries, did not die by the gun. He fell in love with actress and dancer Josephine Sarah 'Sadie' Marcus whom he stole from Sheriff Johnny Behan and later married. They remained together for the rest of Wyatt's life which was a patchwork of highs and lows. Wyatt the professional gambler, yearned for wealth and position, and Sadie who developed a gambling addiction, kept them poor. Wyatt worked in real estate, and saloons in San Diego and refereed prize fights. He joined the Klondike gold rush in Alaska in 1897 where he owned a saloon named The Dexter making a reputed $80,000 that Sadie devoted herself to losing.

Wyatt and Sadie followed other mineral booms in Nevada. Wyatt ran more saloons and was involved in a rigged faro game and arrested, narrowly escaping a jail term. He had a few more short stints as a lawman and spent most of the 1920s dividing his time between a copper mine called the Happy Days and Los Angeles where he became friendly with many people in the movie business.

He told his life story to two separate biographers who both sensationalised it, and he died on 13th January 1929 of chronic cystitis. One of the pall bearers at Wyatt Earp's funeral was silent movie star, William S.Hart; and cowboy turned actor, Tom Mix, was among the mourners.

John Chisum and the Legal Rustle

One of the more interesting stock appropriation cases in America happened to New Mexico cattle baron, John Chisum. Following the Civil War, Chisum, in partnership with two others, went into the meat packing business. Chisum's role in the business was to buy and deliver Texas cattle while his partners Messrs Clark and Wilber ran the processing plant.

Something went wrong in the curing process and the partners found themselves with a mountain of decaying beef and a belly-up venture. This caused Chisum the cattleman to turn his back on the packing business and enter into an agreement with legendary rancher, Charlie Goodnight, to drive herds from Texas to the Pecos on Goodnight's behalf.

Company creditors attempted to sue Chisum for outstanding debts as he was the only one of the partners who had not declared bankruptcy. The cowman laughed at them saying all he could pay in was Texas Longhorns and promptly forgot the whole affair. Like many other entrepreneurial men involved in the post-war cattle boom, Chisum drove a few head of his own along with the Goodnight herds under his care, and soon established a ranch called the Jingle Bob at Bosque Grande on the Pecos River in New Mexico.

When handling cattle in open range conditions, it is a common occurrence for stock belonging to others, to finish up in herds destined for sale. To simplify matters, many ranchers issued a power of attorney to other cattlemen and trail bosses to sell on their behalf. This saved the time-consuming job of cutting the herd on open camps and John Chisum was the holder of several such documents. Chisum may well have been a bit too entrepreneurial for his own good. The story goes he was quite vigilant about caring for his neighbours' charges until sale, but rather haphazard about squaring up for them afterwards.

John Chisum prospered to a high degree. His herds grew to number 100,000 head and ranged for 160 kilometres on either side of the Pecos River. He built a mansion at his South Spring headquarters that could rival any squatter's castle constructed in Australia during the halcyon days of the wool trade, and became known as 'The King of the Pecos'.

The greatest thorns in the side of the Pecos King were rustlers who nipped constantly at the flanks of his kingdom. Chisum and some of his men once pursued and retrieved 1,200 horses driven off by Mexican marauders but the depredations of the brand-blotting brigades, although considerable, could not hold a candle to the loss he sustained in 1876.

The year before, Chisum had faced a lawsuit brought by an aggressive young Las Vegas, New Mexico lawyer named Tom Catron over money unpaid to Texas ranchers who entrusted stock for sale with John Chisum. He was jailed for a few days on a charge of misappropriating cattle but Chisum produced his old powers of attorney and the charges had to be dropped leaving Catron smarting over being made a fool of.

The Dodge City *Globe Newspaper* held a seemingly innocuous news item on 6th April 1876 reporting the departure of one Jesse Evans and fifty cowboys, for the Pecos to round up 20,000 head of cattle purchased from John Chisum by the firm of Hunter and Evans, Kansas City [15]. Before Evans set out for the Pecos country, the senior partner of Hunter and Evans, Robert D.Hunter, had quietly poked around buying up Chisum's old notes from his failed packing venture and various Texas judgements over cattle unpaid for, at 10c on the dollar.

John Chisum was quite unaware of the background machinations going on and eagerly agreed to the sale of such a large number of cattle. He detailed his own men to help Evans' crew round up the stock and some time later, he rode to Las Vegas with Jesse Evans to receive payment.

The first inkling The King of the Pecos got that all was not well, was when Evans led him into the legal office of a smirking Tom Catron. Still seething over Chisum wriggling out of his clutches previously, Catron laid a satchel on the desk and Robert D.Hunter entered the room with the flat statement "Here's your pay, Chisum."

It was probably fortunate for the sharp if not downright shady trio that John Chisum was not carrying a gun when he opened the satchel to find his long abandoned debts staring him in the face. Chisum was a volatile man and the assessment of the pedigrees of Hunter, Evans and Catron that he howled to Las Vegas at large, may well have amused the local population, but it didn't alter the facts.

Following his outburst, Chisum threw himself into the saddle and headed straight for the huge herd being driven out of the country. He met his own men to learn that everyone of the fifty cowboys hired by Hunter and Evans were hard cases recruited in Dodge City for their skill with firearms as much as their ability to drive cattle. Not being in a position to wage war on fifty hand picked gunmen, The King of the Pecos could only grind his teeth and swear as his legally rustled cattle walked away to the east and forever out of his reach.

BLACK, RED & WHITE WARS

Most people in the U.S.A. have heard of 'Crazy Horse'
that Indian folk hero who demolished Custer's force,
'Sitting Bull', the Nez Perce 'Joseph', 'Cochise', 'Geronimo'.
Indigenous Americans whose deeds of long ago
brought a blaze of glory to the red man in his land.
But do we heed our native heroes who trod this southern strand?

Have you heard of 'Nemarluk' who fought by Kimberley lagoons?
How 'Beresford' was vanquished by the fighting Kalkadoons?
Carbines cracked in Lawn Hill Gorge to still Joe Flick's last hope,
and Jimmy Governor's tragic tale that ended at the rope.
Why don't we Australians revere our native sons
who fought to save their birthright and died beneath the guns?.

Jack Drake

Beginnings and Duration

During the era of discovery, powerful nations were planting their flags in new lands as fast as their seagoing navigator/explorers could ride the winds and tides of global expansion.

Following America's discovery by Christopher Columbus in 1492, the French established a foot hold in Canada by 1600 and the English by 1620. From the first landing by Spaniards until the defeat of the northern plains tribes in the 1880s, whites were in conflict with native Americans for almost 400 years.

Apart from being a handy dumping ground for convicts, Australia was ideally suited as a supply base on Southern Trade routes. The Dutch were starting to look hard at '*Terra Australis*' so England decided to get in first. From 1788 until 1928 the 'war that never happened' was going on in some part of the country between whites and Aborigines.

Australian Conflict

When the First Fleet landed it was found that natives in the Sydney area retreated at the sight of firearms[1]. This was most likely prompted by Frenchman, Comte-de-La Perouse, who sailed into Botany Bay at the same time as the first fleet in 1788. He built a stockade there and is reported to have fired upon people of the Bidjigal who occupied the north west coast of Botany Bay [2]. Captain James Cook who landed there in April 1770 made no mention of hostilities in his log but he too may have given natives a demonstration of fire power.

The British settled Australia under the dictum of *Terra Nullius* (a land without people) and it is surprising that these people who officially did not exist, managed to conduct a series of guerrilla wars that lasted well over 100 years. If we believe the way history has been written regarding white settlement of Australia, it appears the indigenous people melted away with only token resistance. This could not be. The Aborigines were owned by the land. They could not leave it because it was the source of their life, religion, culture and their very reason for being.

'Dreaming' is a word that has received a lot of use in recent times but it is the term that comes closest to explaining Aborigines' bond with their territory that can never be truly understood by people outside their culture. Every creek, rock, tree, mountain and animal was an entity in its own right with a soul and a reason for existence. To simply move on in the face of invasion was impossible. Their land was their life and apart from that, crossing into a neighbour's territory uninvited was a death sentence.

Interaction between tribes was a complicated process. Those wishing to cross boundaries for reasons of trade or to discuss matters relevant to both peoples would wait at the border of their territory for an invitation to talk when their signals were acknowledged. Certain times of truce were observed when food supplies such as the triennial Bunya Nut feasts were available and these became festivals where separate peoples would visit and interact. Such occasions were conducted with much ritual and ceremony with times and conditions of entry to the host tribes' lands being strictly observed.

The only individuals, who had relatively free access to the territory of others, were the Corroboree Singers who carried the songs and plays that were the oral history of the Aborigines. They alone could move with few restrictions and singers became the freedom fighters who sought to unite the tribes in the face of white takeover.

Pemulwuy and the Eora Resistance

It did not take long for the Eora people who inhabited the land on the south side of Sydney Harbour, to become disenchanted with whites of the first fleet. They had heard of La Perouse's men and muskets from their neighbours, the Bidjigal and were wary of these strange, pale people who violated their land and over fished their waters.

It was inevitable that tension would boil over into violence and on 20th March 1788, several convicts were wounded in a scuffle when the Eora attacked a work party in the bush. The first white casualty in Australia's war of attrition was convict Peter Burns who fell to Eora spears near Wooloomooloo Bay on 22nd May and the score was soon evened when a tribesman died the following day in a revenge killing at Blackwattle Bay.

An uneasy co-existence between the Aborigines and the convict settlement dragged on with attempts to gain the confidence of the Eora by Governor Phillip. He tried to achieve an inside perspective by capturing Aborigines to learn from them. His first captive Arabanoo died in the smallpox epidemic of 1789 which killed about half of the Eora and is believed by some historians to have been deliberately introduced. His second captive Bennalong, taken in November 1789 stayed with the whites for five months before he escaped and returned to his people.

Phillip saw Bennalong twice in the month that followed and at the second sighting he attempted to approach his former captive who was with a group of warriors at Manly on 7th September 1790. Another tribesman Wileemarin drove a spear into the Governor's shoulder and the natives fled.

Governor Phillip ordered a patrol to make contact with Bennalong. They achieved this through a liking for alcohol he had developed during his captivity. Bennalong and a few followers began mixing with the whites more freely then and became puppets playing the Governor's game in return for grog and other gifts like military uniforms. Not all of the Eora were so easily bribed however, and a warrior, who would become Australia's first noted leader in the war against the whites, arose.

A convict named John McIntire, who was the official hunter of game for the Governor's table, had been guilty of acts of aggression against the Aborigines. He was hunting with a party of two other prisoners and a sergeant of marines when the Eora declared war on 10th December 1790.

The hunting party had camped in the bush and was just beginning to stir that morning when they became aware of approaching warriors. First blood was drawn by the man who would become Australia's first freedom fighter, Pemulwuy, when he leapt over a log and drove a spear into McIntire's chest. The natives were pursued unsuccessfully but the spearing of Governor Phillip's huntsman turned the man who had attempted a peaceful integration of white and Eora, however clumsy, into a tyrant. Phillip ordered the destruction of all Eora weapons, the taking of two male captives, and the execution of ten more Aboriginal men. The dead warriors' heads were to be returned to Sydney.

Phillip's punitive expeditions, which were mistakenly directed against the Bidjigal people, were fruitless as the natives slipped away before their advance, but scattered spearings and an organised raid on an outpost at Rose Bay now Parramatta, showed that Pemulwuy and his guerrillas were active.

Governor Phillip resigned in late 1791 and returned to England taking Bennalong and another Eora man, Yemmerrawannie, with him. Bennalong spent two years mixing in British society where he was regarded as a curiosity of the South Seas. He returned to Australia in 1795 aboard the H.M.S.'*Reliance*' , a detribalised joke who fitted in with neither race. Bennalong died with his Aboriginal people in the bush on 3rd January 1813. Yemmerrawannie never returned to his native land dying in England of lung disease. He was buried in the parish churchyard at Eltham in Kent [3].

An ex-American War of Independence soldier Francis Grose became acting Governor on Phillip's departure and a period of relative quiet occurred until 1793. The second and third fleets were in residence at Sydney Cove by this time and the Eora were probably fearful of the increased numbers. Governor Phillip had attributed seventeen white deaths and woundings to Pemulwuy and his men. The charismatic war leader attained great standing among his people. Believed to be around thirty years old when he began carrying his fight to the whites, he soon earned an almost supernatural standing amongst the Eora and feelings of dread among his white enemies.

The Eora believed he was immune to English bullets and could never be killed or defeated. He showed himself as a noticeably visible target on many occasions returning unscathed from the field of battle and his legend of invincibility grew.

For twelve years Pemulwuy waged war on the white invaders. He began in Governor Phillip's time, was active during the terms of office of Acting Governor Grose and Administrator Captain William Paterson, outlasted Governor John Hunter and was to prove a huge obstacle to Hunter's successor Philip Gidly King. He led raids throughout the Sydney region and was probably allied with the Dharuk people during their Battle of the Hawkesbury in June 1795. He fought a major engagement in what is known as the Battle of Parramatta during the time of Governor Hunter.

In March 1797 Pemulwuy led raids in the Toongabbie area and fought a hit-and-run action against pursuing troops that ended with British forces retreating behind fortifications at Parramatta. The Eora forces attacked and charged time after time. Pemulwuy showed himself in the forefront of the fight eventually going down with seven wounds to the head and body. Half the war party were mown down in that ill advised frontal assault on entrenched riflemen and the Eora were forced to retreat leaving their leader amongst the dead.

Pemulwuy was taken to a hospital in Parramatta where he lay chained waiting for the death his tribe believe he had suffered and his enemies considered inevitable. Against all odds, the war leader recovered and feigning weakness, he caused his guards to be inattentive and managed to escape in the night with a leg iron still in place.

The Eora rejoiced in their Chieftain's return and his legend of invulnerability grew with his return from the dead. After regaining his strength, Pemulwuy began raiding again and using a new tactic. He admitted escaped convicts to his ranks to fight the British Administration. He was a living legend to white and black alike by this stage in the Sydney region, and when Governor Hunter left for England in 1800 and Phillip Gidley King took his place, Pemulwuy was the number one challenge for the new Governor.

King conscripted settlers in the fight against the guerrilla leader and set a reward on Pemulwuy's head. Twenty gallons of spirits and two suits of clothes for a freeman, a ticket-of-leave for convicts, and a free pardon for time-expired prisoners, were the baits dangled for Pemulwuy 'dead or alive'.

Members of the Eora were invited to turn traitor and 'give up' their leader who was held in godlike status among them. The reward King offered them was to be 're-admitted to our friendship'. Big deal! Needless to say none of Pemulwuy's people sought the reward.

Two English settlers, whom records have avoided naming, brought down the guerrilla leader in late 1802. Pemulwuy's head was cut off, pickled and sent to botanist Sir Joseph Banks as a scientific specimen along with Governor King's confident assurance that the Eora were now very friendly.

It was not to be. Pemulwuy's son Tedbury, who had fought at his sire's side, took up the cause and harassed white settlers for another eight years. Tedbury carried on his father's practice of using escaped convicts against their former masters and also began using firearms in a small way.

The Eora were a doomed people however. Years of war and white diseases had whittled their numbers lower and lower. When Tedbury was shot dead in 1810 by an English settler named E.Lutterill, all resistance ended and the fighting Eora Nation ceased to be.

The War for the Bathurst Plains

Following the discovery of a way through the Blue Mountains to Sydney's west by Gregory Blaxland, William Lawson and William Charles Wentworth in 1813, settlers began moving into the promising pastoral lands known as the Bathurst Plains.

In the style that would come to be typical of Aboriginal reaction to white advancement, the Wiradjuri people of the region made themselves scarce. When finally contacted, however, they proved friendly to the newcomers. When Governor Lachlan Macquarie first visited the area in May 1815, his camp was visited by a delegation of Waradjuri warriors headed by a man who would eventually dash the Governor's hopes of a violence-free takeover of the new land.

Windradyne was a tall, fine looking, muscular man who accepted Macquarie's offering of tomahawks and yellow cloth with the dignity of a born statesman and returned the gesture by presenting the Governor with a fine possum skin cloak.

By July that year, cattle were grazing the Wiradjuri hunting grounds but Windradyne and his people maintained friendly relations with the whites for seven years until Macquarie was replaced by Governor Thomas Brisbane.

Brisbane was of a scientific bent being far more interested in astronomy and maintaining a genteel lifestyle than attending to the Colony's affairs. While the new Governor occupied himself with his household, social pursuits, hobbies and wine cellar, the Surveyor General John Oxley and Colonial Secretary Major Frederick Goulburn were as good as 'running the show', and allowed almost unrestricted occupation of Wiradjuri land [4].

The first attacks on settlers by Windradyne and his men, began west of the fledgling town of Bathurst in 1822 at *Swallow Creek* and Mudgee. They continued raiding camps and spearing stock, causing the abandonment of the *Swallow Creek* Government Station in November 1823.

Swallow Creek was re-occupied by 19th March 1824. When about sixty warriors raided a hut there, Privates Softly and Epslom of the 2nd Somersetshire Regiment, who were at the neighbouring *Wylde's* Station, investigated with *Wylde's* Overseer Andrew Dunn and Stockman Michael McKegney.

The soldiers and stockmen surprised some warriors in the hut killing two and capturing three more. The Wiradjuri retaliated killing seven stockmen in a series of attacks during May, and a vigilante raid by civilians killed three more tribesmen [5]. John Johnston, John Nicholson, Henry Castles, John Crear and William Clark were found not guilty of manslaughter by Saxe Gannister, the New South Wales Attorney General, and a week later on 14th August, Governor Brisbane declared martial law west of the Blue Mountains.

Brisbane's Proclamation gave Bathurst's Commandant James Thomas Morisset almost unlimited powers to suppress the Wiradjuri and he obtained another detachment of the 2nd Somersetshires bringing his force to seventy-five soldiers.

A series of punitive raids that have become known as the Battle of Bathurst, caused the surrender of the Wiradjuri but not before twenty-two white settlers were killed. The number of Waradjuri deaths is not available, but all sources seem to agree it was considerably more than those of their conquerors.

Windradyne and 260 of his people journeyed to Parramatta where they met Governor Brisbane at midday on 28th December 1824. The Governor accepted the Wiradjuri surrender and lifted martial law. After a year of war the Bathurst district was in the hands of its invaders.

In May 1825, Governor Brisbane was sacked and he returned to England to follow his scientific pursuits. Commandant James Morisset also went to England but returned becoming Superintendent of Police and then Commandant of Norfolk Island Penal Colony until his retirement in 1834. His son later followed in his footsteps becoming an Inspector and eventually Commandant of that murderous force, the Queensland Native Mounted Police.

Windradyne kept the terms of his surrender and tried to ease his people into the new order. He died on 21st March 1829 when a knee wound received in a tribal dispute turned gangrenous.

The Tasmanian Annihilation

Historians who attempt to minimise the effects of white colonisation on Aboriginal society, have claimed genocide was not practised in Tasmania. Their arguments seem to focus on academic discussion about the exact meaning of the term genocide, but the fact remains in the first thirty years of white settlement, Tasmania's native race was reduced to around 200 confused, broken people who were herded onto offshore islands where they dwindled and died out.

The history of Aboriginal dispossession by whites in Australia, is being written by two distinct camps. The so called 'Black Armband' historians dwell on massacre and atrocity and never miss a chance to paint lurid pictures of indigenous misery and suffering. However, they have ample evidence to back their claims and only a fool would give credence to the view that this country was ceded peacefully, which is stubbornly adhered to by officialdom.

This constant sparring between academics is purely academic. If some of the 'Black Armband' Brigade are a bit 'over the top' their opponents are certainly 'under the thumb' of the version it has suited white history to record. Tasmania's settlement has been cited around the world as a prime example of genocide and all the available evidence points that way. We can talk technicalities until the cows come home, but we cannot escape the fact that the original, full-blooded inhabitants of Tasmania will never come home.

Van Diemans Land, as it was known then, was first occupied by whites in September, 1803 when forty-nine people formed an embryo colony at Risdon Cove on the Derwent River near where Hobart now stands. In February of 1804 they were joined by the remnant of the failed Port Phillip settlement. (chapter 1 volume 1). Van Diemans Land was first used as a dumping ground for incorrigible convicts from Sydney Harbour. It became an independent colony in 1825 and the island has a bloody history as a convict hell as well as for its infamous treatment of Aborigines.

Van Diemans Land natives had some contact with whites before settlement, not all of it friendly. When Abel Tasman landed in 2nd December 1642, he met none of the island's inhabitants but the second European we know of who visited its shores engaged in a brief conflict.

Marion du Fresne arrived on 5th March 1772 and a few locals warned him off by chucking some well-aimed rocks. The Frenchman was injured by his first contact with a chunk of Tasmania and he ordered a volley which killed an Aborigine and wounded several more [6].

The Derwent River settlement was established and conflict followed soon after. A large party of Aborigines spread out in a crescent shape, moved down the hillside behind the village towards the beach on the morning of 3rd May 1804. Lieutenant William Moore of the New South Wales Rum Corps saw them coming and decided attack was imminent.

What was really taking place was a kangaroo drive where large groups of people would sweep an area driving animals before them to the water then dispatch them with clubs as they tried to break back through the line. To be fair to the Derwent settlers, 300 odd natives spread out, advancing armed with waddies that could have been mistaken for spears, would have looked very threatening. The Lieutenant ordered his troops to fire when the natives came within range. Quite a few Aborigines fell dead and wounded. Moore officially admitted three deaths. The rest abandoned the hunt, fleeing for the bush with alacrity, and who could blame them.

Lieutenant Moore's attack may have been motivated by malice but it is more likely the first clash between Tasmanian settlers and the island's indigenous inhabitants was the result of a misunderstanding. There did not seem to be much misunderstanding about events that followed however.

True to the Aborigines' first reaction to white occupation, indigenous Tasmanians attempted to remain peaceful and the first twenty years of settlement was relatively quiet. When practically all of the low lying, arable land on the mountainous island had been claimed and when atrocities committed by settlers and convict 'bolters' had become too much to bear, the natives finally said "Enough!"

The war in Tasmania began in earnest in 1824. The natives, in no position to launch frontal assaults, used guerrilla tactics attacking isolated farms and particularly killing livestock wholesale in an attempt to destroy the settlers' livelihood. In the dryer summer months, they employed fire to burn crops and buildings and for seven years conflict raged across the picturesque landscape of the island state.

In writings left by George Augustus Robinson, the only member of the colony who spoke the tongue of the Big River and Oyster Bay peoples, he spoke of white retaliation by men like settler William Gunshannon who killed right and left as the whim took him.

The warfare peaked in 1830 with 222 separate engagements that year, and when William Gunshannon fell to native spears, Robinson commented that "The Justice of God was apparent."

Like Tedbury of Sydney's Eora, Tasmanian Aborigines learned to use firearms captured in raids. In the fashion of indigenous people newly introduced to guns worldwide, they made the mistake of thinking the harder the trigger was pulled, the further and faster the bullets would fly. This was a logical assumption for warriors trained in the use of spears. Despite the fact their accuracy was fairly questionable, the original Tasmanians were an inventive race. They made replacement flints for muskets and fired upon whites but only one settler died as a result when a group of Big River warriors attacked farm overseer James Parker near Port Sorell and shot him to death with his own gun at close range in September of 1831 [7].

At first Governor George Arthur attempted the same tactics he had used to defeat bushranging gangs of escaped convicts that infested Tasmania's wild lands in the previous twenty years. He ordered troops into the field against the Aborigines in 1826 and over 200 soldiers were active on the frontier. Arthur's soldiers enjoyed little success. Their bright red uniform coats made them extremely conspicuous. By Aboriginal standards, they were ridiculously noisy as they thrashed their way through the dense bush. The odour of spirits and tobacco they emitted, made them easily detectable to keen Aboriginal noses and their cumbersome Brown Bess Muskets were constantly getting hung up in trees and undergrowth.

Rather than dealing with the problem, the presence of soldiers seemed to cause an increase in raids. Governor Arthur at his wit's end, declared martial law over parts of the colony on 15th April 1828 and made it illegal for natives to enter settled districts. Tasmania was in an uproar and at a meeting of the Land Executive Council on 27th August 1830 it was decided to raise civilian militia units to aid the military.

Governor Arthur commissioned Major Sholto Douglas of the 63rd Regiment as the Civil and Military Commander and a bold plan was hatched. A force of 2,200 men was raised to sweep the settled districts in an extended skirmish line and drive the troublesome traditional owners onto the Forestier and Tasman Peninsulas which would then become a reservation for them.

On 1st October 1830 the Governor extended martial law to the whole of Van Dieman's land and the 'Black Line' began its advance on an almost 200 kilometre front from the north and west. Troops and civilians battled up and down hills hacking their way through the bush and halting from time to time to let laggards catch up. By 24th October, the northern and western contingents joined and concentrated on a fifty kilometre front when the heavens broke and rain bucketed down causing the advance to halt as men huddled under whatever shelter they could find.

During the stoppage, patrols went forward to report on Aboriginal activity and a civilian party led by Edward Walpole skirmished with a group of Aborigines killing two and capturing a man named Ronekeennarener and a boy named Tremebonenerp. The rain came on in earnest forcing a delay of three weeks and making movement of supplies to the line a nightmare.

Arthur sent extra men and dogs as sentries. Obstacles and rough fences were constructed to try and hold the line against the native hordes believed to be contained in front of it. Extra rations of salt and tobacco were ordered and cattle and sheep for slaughter were sent in to keep morale up. Eventually the weather dried up enough to get the Black Line moving again.

Surprisingly morale remained high. Only two convicts had their tickets-of-leave revoked for desertion and one settler John Crouch was banished for 'endeavouring to create a spirit of discontent'[8]. The final advance to the neck of Forestier Peninsula began on 17th November and a week later on the 24th, the operation ended.

Not one Aborigine had been driven to the proposed reservation. Two deaths and two captives were the total result of an expedition that cost £30,000. No doubt the Aborigines, who had slipped through the cordon with ridiculous ease, sat on the Tasmanian heights and watched with wry amusement as the Black Line was disbanded. They may even have laughed a bit, but the loss of lifestyle and loved ones probably made this unlikely.

The Big River and Oyster Bay people were decimated, and Governor Arthur's next solution was to relocate the remnants of the tribes on islands in Bass Straight.

Sixty of them were removed to Gun Carriage Island where many died of homesickness and starvation. In late 1831 the remnants were transferred to Flinders Island along with others rounded up on the mainland and this practice continued until 1847 when the Flinders Island settlement was closed and forty-four people were returned to a reserve at Oyster Bay [9]. A man, William Lanny and a woman, Truganini, have been claimed as the last of the full-blooded, Tasmanian Aborigines. Lanny died on 3rd March 1869 and Truganini followed on 8th May 1876 but people of mixed white and dark descent live in Tasmania today who can still lay claim to Tasmanian Aboriginal ancestry.

Tasmanian Aborigines were of a different genetic make up to tribes of mainland Australia. They are believed to be of an older race present on the continent when Aboriginal people entered by the land bridge to Asia at least 60,000 years ago. In appearance they looked very similar to the remnant of New Zealand's Moriori race who were killed out by Maoris who entered the country by sea from islands they had overpopulated in Polynesia in the 14th Century [10].

The last people of Moriori blood died out in the Chatham Islands off the east coast of New Zealand in the mid 20th Century and from photographs, they look very similar types to Tasmania's indigenous people. It could be that both races were part of a similar gene pool.

Myall Creek

One of the most significant events in the wars between black and white in Australia occurred at *Myall Creek* Station west of Inverell on the New England Tablelands, New South Wales in June 1838.

Publicity about ill treatment of Aborigines had spread to Britain prompting Lord Glenelg, Secretary for the Colonies, to send a directive to Governor George Gipps in 1837 that native people were to be treated as British subjects.[11] A few efforts had been made previously to discipline Europeans who committed outrages but white courts had little sympathy for the natives they saw as less than human, and no convictions resulted.

Squatter Henry Dangar's men drove a herd of cattle to *Myall Creek* on the Big River Catchment (Gwydir) in 1837. Dangar was an influential man in N.S.W. with extensive squatting interests - a land taker who saw the new country as his ticket to riches [12].

At first Dangar's stockmen formed cordial relations with the Weraeria people of the region. The tribe camped near the new station huts interacting with the newcomers quite harmoniously. Everything changed on the evening of 9th June 1838 when a group of horsemen from other stations nearby, heavily armed with guns and swords, rode up to the huts where convict workers Charlie Kilmeister, George Anderson and thirty-three Aborigines waited. The men claimed to be hunting stock-spearing natives.

The leader of the party was John Russell transported for theft and now a ticket-of-leave man working as overseer on nearby *Bengari* Station. Russell's second-in-command was another transported felon, George Palliser and the rest of the party was made up of assigned convicts, Ned Foley, John Johnstone an Afro-American, James Hawkins, James Oates, Charles Toulouse, Australian born John Flemming plus three other men Jem Lamb, James Parry and John Blake [13].

The posse fastened all but five of the Aborigines, who had crowded into one of the huts in alarm, together with a long, tether rope while Irishman Ned Foly guarded the door with a cocked gun. The bound and weeping natives were marched west towards the hills and an eerie silence descended as they disappeared from sight.

Charlie Kilmeister got his horse, sword and gun and cantered off to join the grim procession but Anderson wanted nothing to do with the menacing mob. He stayed by the huts terrified for his own safety and that of the few Aborigines the marauders had left. Yintayintan, an Aboriginal worker known to the whites as Davy, was part of the station staff and one of those not taken. After sunset he followed the track of the kidnappers to discover a gruesome sight. Near a recently built stockyard, twenty-eight bodies lay hacked to pieces by swords because the murderers had decided it was cheaper than shooting.

One attractive young maiden was spared for Russell and his fellow fiends to rape repeatedly with the blood of her people on their hands. The unfortunate who would probably rather have been hacked to death with her family, was then dragged along as the party sought further bloodshed.

Ten of the Kwiamble people were known to be cutting sheets of bark for a squatter in the area, and they were the next targets. Unsuccessful in their quest for the bark cutters the group rode to Dight's station *Keriengobeldi* the next morning, where they ate and rested with hutkeeper John Bates and boasted of their deeds. The young woman was absent by then, her captors obviously having disposed of her along the way.

That night the murderers showed up at Eaton's *Biniguy* property. Mr. Eaton had sheltered and fed the bark cutters, who were aware by now they were being hunted, but in the face of the armed party, all he could do was advise them to flee into the hills. The posse pursued them and three men and a child were shot, the rest escaping. After this last outrage, the killers went their separate ways each man returning to his home station. Kilmeister returned to Myall Creek where he spent time burning bodies and clearing the evidence up at the massacre site.

News of the Myall Creek killings rebounded around the district and when Henry Dangar's manager, William Hobbs, returned from a short droving trip, he reported the matter to Muswellbrook Police Magistrate, Edward Denny Day. Day reported to Governor Gipps who determined he would carry out his directive to treat Aborigines as subjects of the Crown and prosecute the offenders.

Warrants were issued for the twelve perpetrators charging them with the killing of 'Daddy' a gentle giant of a man who fell victim of the massacre. Somehow native born John Flemming got wind of his impending arrest. He took to the saddle and by riding day and night acquiring fresh horses where he could, the fugitive reached Sydney. He is reputed to have taken ship for Tasmania and was never seen again.

Those were very different times and an incredibly callous attitude by today's standards prevailed against Aborigines. A group of graziers met at Patrick Plains (now Singleton) and pledged £300 for the colony's best lawyers to defend the accused. Squatter Henry Dangar contributed to the fund and sacked William Hobbs his manager [14].

The eleven stockmen from the Big River (Gwydir) country were tried in Sydney Supreme Court on 15th November 1838 for the killing of 'Daddy'. In the face of overwhelming evidence a jury of white colonists declared them 'Not Guilty'. The men emerged triumphant only to receive a nasty surprise. They were re-arrested on a charge of killing an un-named child and thrown back in jail.

Eleven days later, Kilmeister, Foley, Oates, Russell, Hawkins, Johnstone and Parry stood in the dock to hear a guilty verdict handed down. The other four men Lamb, Palliser, Blake and Toulouse were not tried a second time and were released. The seven *Myall Creek* murderers 'sprung the trap' in Sydney on 18th December 1838 and although British justice was seen to be done, the hangings had the effect of driving persecution of indigenous people underground.

No more would brutal graziers and their henchmen brag of killings and the savage dogs they kept to ' pull down niggers'. Reprisals were not going to stop but now they would be carried out furtively and their perpetrators would adhere to the law of silence under a bond prompted by the threat of the noose.

Myall Creek was a massacre in the true sense of the word. The Weraerai, Kamilaroi and Kwaimbul Aborigines who occupied the Big River country were an extremely peaceful people. They had not yet begun to fight back and were living in harmony with settlers when the events at Dangar's Station took place.

Bielbah the Freedom Fighter.

According to the testimony of two convict bolters Davis and Bracewell, from the Moreton Bay Penal Colony, who lived among the Whaka Whaka of the Wide Bay area for fourteen years, a tall, well formed, young warrior named Bielbah was actively trying to unite the tribes. The escaped convicts had attended the triennial Festival in the Bunya Mountains when Bunya Pine kernels ripened from January to March of 1841. They made their way back to Brisbane in 1842 and white Australians first heard of Bielbah [15].

Bielbah is believed to be of the Gungabula people who inhabited the country around the south-east end of Central Queensland's Carnarvon Ranges. He was more than likely a corroborree singer as he seems to have moved unhindered among the tribes.

The 1844 Festival of the Bunya fruits hosted by the Kaiabara people would appear to be when the decision was taken to resist white invasion. It is easy to imagine Bielbah the warrior poet prowling corroboree grounds in the firelight, backlit by leaping sparks, his savage features contorted with hatred as he whipped the tribes into a frenzy to the click of the rhythm sticks and the drone of the didgeridoo. He must have been a powerful orator and the Bunya Festivals would have given him a perfect stage to incite the tribes to war.

Native Australian people came from as far south as the Clarence River district of New South Wales, Gladstone in the north, off coast islands like Fraser and Moreton and as far west as Roma to the Bunya Festivals. They made the trek every three years when traditional boundaries were relaxed to allow peoples from hundreds of miles away to share the bounty of the Bunyas [16].

Much has been learned of tribal migrations to the Bunya grounds called Bunnia-Bunnia by the Kaiabara from rock paintings discovered in the Maidenwell district. Carbon dating of campsites there, prove the triennial migrations had been occurring for over 4,000 years [17].

Bielbah launched his first offensive against the whites in 1848 when he organised and led a series of raids on *Mt. Abundance* Station taken up by Alan MacPherson on Muckadilla Creek in the Roma district. In the first strike, a shepherd named Lowe was captured and tied to a post. Bielbah and his men used him for target practice hurling twelve spears through his body. Lowe's flock of 2,000 in-lamb ewes were then driven off.

Not long afterwards, a hut keeper named Gore was skewered to his hut door with a spear causing thirteen other MacPherson employees to take to their heels leaving Bielbah and his raiders in possession of the station. At the time, Alan MacPherson was returning to *Mt. Abundance* with a load of supplies. He found his terrified men in a camp on the Balonne River where he managed to shame and cajole them into returning with him. All was quiet on MacPherson's return, Bielbah and his men having taken to the hills to feast on MacPherson mutton [18].

It did not stay quiet for long though. Bielbah raided the station time after time. Alan MacPherson was a dour Scot and he hung on with the tenacity of his race. He tried to fight the raiders, but they struck, killing shepherds and sheep, to vanish again like smoke in the trees. During 1848 nine of MacPherson's men fell to the spears of Bielbah and his followers as well as most of his flocks.

The following year, Alex Blyth of the neighbouring *Blythdale* property was speared and two of his men killed, forcing Blyth to abandon the fledgling property [19]. The Roma district was in an uproar in 1849. Bielbah and his men killed two men at *Dulacca*, three more at *Tieryboo* and a Mr. Jones died at Wallan where a store was burnt as well. Alan MacPherson finally threw in the towel bitterly admitting defeat and drove his remaining sheep east out of his enemy's territory.

In September of 1849, Bielbah took on a contingent of Frederick 'Filibuster' Walker's Queensland Native Mounted Police (chapter 3 volume 1) sent up from *Callandoon* near Goondiwindi to hunt him down. As usual the Native Police were fairly thorough, killing several of Bielbah's warriors but their primary quarry escaped and seemed to be content to disband his men and rest on his laurels for a while [20].

Five years later, he resurfaced with a vengeance. He led another raid at *Dulacca* Station where a man named Kettle died and then Bielbah shifted operations to the Dawson country where he allied himself with the Jiman Aborigines.

A squatter named MacLaren and two of his men were his first victims in the Dawson River valley. This was followed by other killings and stock spearings, but it all was small beer compared to the raid he led three years later.

Oral Aboriginal history records the Bunya Festival of 1857 as being the place where the decision was made to wage all out war on settlers in the Dawson [21]. The Jiman were a warlike people who dominated their more peaceful neighbours and they would have welcomed a man like Bielbah to their cause.

A squad of Queensland Native Mounted Police was stationed in the Dawson by then, with troopers recruited from tribes in southern New South Wales, who were not bound by local tribal strictures. They and their white officers had outraged the Jiman by casually using their women. Under Aboriginal law marriage was arranged and it was a deadly sin, punishable by death, to have sexual contact with a member of the wrong totem group. Because of the isolation of their lifestyle, arranged marriage was vital to native Australians as a safeguard against inbreeding. Modern genetic engineers have never managed to improve on the totemic system worked out by this 'primitive'

Jiman tribal lands today in the immediate vicinity of *Hornetbank* Station.

Photo courtesy Jack Drake collection

race. The unsanctioned use of their women, consensual or not, was regarded as rape by the Jiman.

As well as the Native Police making free with their women, three brothers, the Frasers, who managed *Hornetbank* Station for grazier Andrew Scott, also had a reputation locally as being 'famous for the young gins'. This was probably the reason for *Hornetbank* being selected as the opening engagement in the Dawson War.

William Fraser had taken over the management of the station at age twenty-two following the death of his father John. He was assisted by his brothers John and David. Also living on the property were the boys' mother, Martha, her daughters Elizabeth aged 19, Mary who was 11, Jane 9, Charlotte 3 and two younger sons Sylvester 14 and James 6.

Martha Fraser was very kind to the Jiman people she came in contact with, but she was well aware of her older sons' shortcomings and had repeatedly asked Lieutenant Francis Nichol of the Native Police to chastise her sons for 'forcibly taking the young maidens', saying she expected harm to come of it.

Martha Fraser was right. At first light on 27th October 1857 Bielbah led the warriors of the Jiman on a dawn raid that would shock the nation. The wily Bielbah had got to an Aboriginal servant of the Fraser's named 'Left Handed Bally' sometimes called 'Boney'. Before dawn Bally (pronounced Bawly) had crept out and brained the dogs with a waddy to prevent them giving tongue as the marauders crept up on the sleeping station.

Fourteen year old Sylvester Fraser who was commonly known as 'West' awoke to hear Aborigines talking in the house. A few seconds later, he was struck down by war club blows to the head and body. He fell between the bed and the wall and was left there for dead as raiders swarmed through the station buildings.

James de Lacy Neagle, the Fraser children's tutor, was one of the first to die with John, David and six year old James Fraser following. William, the eldest was absent freighting wool to Ipswich (which was known as Limestone Hill then) and stores back. Martha Fraser and her daughters were taken outside and West Fraser regained consciousness to hear the warriors beat his mother and sisters to death after raping Martha and the two older girls Elizabeth and Mary.

This was one of the few times white women were raped in Aboriginal raids in Australia. Tribal law operated under 'an eye for an eye' system and it can be assumed that the outrage against the Fraser women was in direct reprisal for the Fraser sons' rape of Jiman women.

Full light came and the blacks discovered two station hands, Shepherd John Newman and a German hut keeper named Bernagle, cowering in hiding. These unfortunates joined the rest of the dead after a short chase. The tally now stood at eleven white deaths.

The attackers left and young West Fraser crawled out of concealment. Under severe trauma with a gaping scalp wound, he walked ten nerve-wracking miles to neighbouring *Eurombah* Station clutching a cocked revolver, terrified that Bielbah and his followers would burst out of the scrub at any second.

Eurombah was shearing and a party of men returned to *Hornetbank* where they buried the victims. Following the burial of his family, fourteen year old West Fraser began an epic ride that is described in chapter 6. When West arrived back with his brother a week later, William Fraser stood over the burial place of his family with an upraised hatchet in his hand and swore a terrible oath, "I swear by Almighty God I will never rest until I have sunk this into the head of the Black responsible". William Fraser apparently gave no thought to the fact that he was one of the main causes of his family's fate and in the months and years that followed, Fraser lived for little but killing Aborigines. His story is told later in this chapter.

Bielbah continued his vendetta against the whites in the wake of the *Hornetbank* massacre and his death has been officially recorded and blood money paid for his capture. A pioneer woman of the Roma district, Mrs. Hannah Cox of *Gubberamunda* Station informed on a man she named 'Bilbo' who was wanted in connection with the Fraser murders. He was shot by a Native Policeman in 1860. Police apparently learned of his plan to attend a corroboree at Cattle Creek near *Mt.Abundance* and shot him as he neared the site. Mrs. Cox received reward money for informing the police of 'Bilbo's' movements claiming to have learned of them from Aboriginal women [22].

The question is, was 'Bilbo' in fact Bielbah? Pollet Cardew the owner of *Eurombah* Station, and other witnesses, claimed Bielbah was present assisting the Gayiri tribe when they massacred the Wills family at *Cullin-La-Ringo* on the Nogoa watershed north of Springsure on 17th October 1861. The Wills massacre was the largest single killing of whites by Aborigines in Australia's history. Horatio Wills and sixteen members of his staff and their families, died eleven days after their arrival at *Cullin-La-Ringo* which Wills had planned to turn into the show station of Queensland [23].

Pollet Cardew's report that Bielbah was not shot at Cattle Creek is right. White people may have believed he died under Native Police guns, but Aboriginal historians know better. The cunning freedom fighter did not go nearly as far as his enemies supposed. He took a different name and assumed a low profile at a native camp in the Auburn River area just to the east of the Dawson near Mundubbera [24]. It is believed Native Police Troopers who by now had a huge respect for Bielbah as a fighting man, deliberately identified the wrong man at Cattle Creek.

Bielbah lived many years in the Auburn River country and as an old man around the turn of the 20th century, he came to the Aboriginal camp near Murgon, Queensland that would officially become the Cherbourg Aboriginal Settlement in 1904. The old warrior passed away in 1904 or 1905 and is buried in an unnamed grave like the bulk of those who sleep in the Old Cherbourg Cemetery.

Bielbah was seen as a murderous fiend by whites and he has been almost ignored by recorders of history. However, he can also be regarded as a freedom fighter who

Cherbourg's old cemetery. The final resting place of Bielbah, the Aboriginal Freedom Fighter.

Photo courtesy Jack Drake Collection

rose among the tribes of central Queensland fighting an heroic if futile war against invading forces who over ran his sacred homeland.

Reprisals against the Jiman people caused considerably more Aboriginal deaths than the whites who died at *Hornetbank* and other skirmishes in the Dawson River country. The officially estimated death toll by Native Mounted Police, irregular vigilante groups and the two surviving Fraser brothers, was given as around 150. The reality is believed to be many more.

Jiman tribal markings cut into the chests of warriors at their manhood initiations were crescent shaped scars. To be the bearer of these curved cicatrices was a death sentence in the Dawson after *Hornetbank*. The Jiman never recovered as a people and their remnants scattered. The last trace of their language died out at Cherbourg Aboriginal Settlement during the 1940s.

Cherbourg was used as a dumping ground for people of many tribes, a lot coming from northern regions. Fifty-three separate tribal groups were herded together there. The settlement was an attempt to break down traditional tribal culture by deliberately merging so many separate peoples [25]. In one way it succeeded as individual cultures were lost in the grouping of many tribes. The Jiman culture became totally forgotten as did all the separate legends and corroborees, but a new order emerged like a Phoenix from the ashes of those destroyed.

The remnants of all the cultures banded together to form the 'Barambah People'. Cherbourg residents created their own corroborees with dances to re-create the birth of a new Aboriginal culture. The fighting spirit of Bielbah has survived with Barambah Aborigines going from strength to strength.

In 2004 Chris Sarra, principal of the Cherbourg State School, was honoured as Queenslander of the Year for his efforts in getting the new generation to understand they need to be educated to gain recognition and respect. Perhaps the old freedom fighter would have liked that.

Fraser the Avenger

When William Fraser swore an oath of vengeance over the graves of his mother and siblings, it began a dark legend of a man obsessed with hatred. Wildly exaggerated stories abound concerning Fraser's depredations, but it is certain that he and his younger brother West (Sylvester) single-mindedly pursued and killed Aborigines for at least five months after the *Hornetbank* massacre. It is also certain that as long as he lived, William Fraser bore a hatred of indigenous Australians and would kill them at any opportunity.

There is a story that Fraser was given a permit by the Government of the time to hunt down Aborigines but this is untrue. He was however given an unofficial assurance by police that no action would be taken about his operations or those of his brother's while they 'got square'.

When William swore to get the 'black responsible' the two obvious choices were Bielbah or Left Handed Bally or Boney who had turned against the family. Fraser later claimed to have carried out his promise and a spot on Juandah Creek known as Fraser's Revenge is where legend says it occurred. A large killing of Aborigines did happen at a lagoon on *Juandah* Station during the *Hornetbank* reprisals.

In the memoirs of a Taroom man born in 1884, he recalls visiting *Juandah* as a boy and stockmen showing him the spot where natives had fled across the lagoon with Fraser shooting into every head that appeared above the water [26].

If the 'black responsible' was Bielbah or Bally, Fraser did not get either. Bally was a southern Aborigine brought to *Hornetbank* by owner Andrew Scott. Following his betrayal of the Frasers, he fled into the Carnarvon Ranges to the northwest and joined a group of about thirty renegades who hid out there, wanted for murder and stock spearing.

In 1860 Charles Coxon, a station owner in the Condamine district, learned from station natives that the gang was planning a raid. Coxon advised the Native Police who arrived just as the raiders ran up to the station. Recent heavy rain made the black soil country nearly impassable and the police were at a severe disadvantage with their horses floundering in the mud.

Second Lieutenant Carr attempted to parlay with the invaders but was met by a shower of spears that wounded several troopers before the police opened fire. Some of the renegades went down but 'Left Handed Bally' urged his mates to keep up the attack. He had seen Carr handing cartridges from his own pouch to his troopers and knew the police were short of ammunition, but they must have had enough for when the fight ended fifteen Aborigines lay dead, Bally among them.

Carr gave evidence to an 1861 select-committee inquiry stating on oath there was no doubt Bally died in the raid on Coxon's station [27]. This would appear credible evidence that William Fraser did not kill Bally considering many whites were familiar with him during his time at *Hornetbank*.

If he did not get Bielbah or Left Handed Bally, William got plenty of others. He once met up with a man travelling in company with a young Aborigine who had been raised as a member of his family. Fraser announced he was going to 'shoot that Black'. The young man's unarmed guardian pleaded for the youth's life, saying he had nothing to do with the Dawson country, but Fraser was implacable threatening to shoot him too if he interfered.

The most common tale of William Fraser's vengeance is claimed to have happened in Rockhampton, Ipswich, Emerald, Springsure, Dalby, Drayton and Toowoomba as well as a few other Queensland towns. The most likely locations are either Dalby, Toowoomba or Drayton in the spring of 1858 but regardless of confusion about where it happened, it is too persistent a legend for there to be no truth to it.

Fraser came riding down the main street of the town in question and saw an Aboriginal women wearing a dress he believed had belonged to his mother. Without a second thought, he stepped off his horse, whipped a revolver from his holster and shot the woman dead.

This and another incident where Fraser openly shot an Aboriginal strapper at a Toowoomba Race Meeting, are believed to be the cause of Police warning him enough was enough. It seems he did not take the warning seriously though.

In the memoirs of Rose Roseby of *Gubberamunda* Station near Roma, she recalls her father and another man standing Fraser off with rifles when he arrived at the station intent on shooting Aborigines in 1860 and both Fraser brothers were present at reprisals for the Wills massacre at *Cullen-la-Ringo* four years after *Hornetbank*.

Legend has it that William Fraser personally shot one hundred natives but this is probably exaggerated. There is no doubt however, that he extracted a bloody retribution for the killing of his family and became a murderous psychopath in the process.

The Fraser brothers did not succeed to any degree in their frontier occupations. William tried his hand as a squatter and when his station failed, joined the Queensland Native Mounted Police which would have caused more Aboriginal deaths. In the late 1880s he moved to Roma and became a drover [28]. He never lost his hatred of Aborigines and as an old, white haired man cooking in his son-in-law's stockcamp

Fraser gave an Aboriginal stockman a dirty look and told a visiting cattle buyer named Handly, that he'd "just as soon shoot the black bastard as cook his tucker."

William died in Mitchell Hospital of bowel cancer on 2nd November 1914 and is buried in Roma.

West Fraser drifted away from his brother and became a carrier first in north Queensland's sugar industry and then around Normanton in the Gulf country. He was prone to periods of insanity and was admitted to Woogaroo Asylum at Wacol, a suburb of Brisbane, on 23rd August 1898. He died on 22nd June 1899 and lies in an unmarked grave in the hospital cemetery. Woogaroo is now known as Wolston Park Hospital.

The Fighting Kalkadoons

The tribe most noted for its fighting ability and unyielding resistance to white settlement in Australia, was the Kalkadoons (Kalkadunje). Tall, well-built warriors, they lived like black liege lords in the rugged hills of Western Queensland around what is now the Cloncurry and Mt. Isa region. Their neighbours the Pitta Pitta, Wunumara, Undekerebina, Workoboongo, Mayithakurti and Workia tribes were thoroughly cowed by these tough highlanders and the Kalkadoons extracted tribute from them in lives, supplies and women whenever they felt like it.

The Kalkadoons had their first look at a white man when Burke and Wills' ill fated exploring party crossed a corner of their land en route to the Gulf of Carpentaria in early 1861. Burke and Wills' rather sketchy records list no meeting with tribesmen when they discovered and named the Cloncurry River but the Kalkadoons' oral history later noted, says they were sighted.

Elderly Kalkadoons in the 1900s remembered the first sighting of these strange travellers. Young men of the tribe spoiling for a bit of fun were all for attacking, but thought better of it when confronted by the fearsome monsters who accompanied the explorers. In light of subsequent depredations on whites by the Kalkadoons, Burke and Wills would appear to have been very fortunate they had camels as pack animals [29].

The next whites in the area were three parties sent to search for the missing explorers. They were led by William Landsborough, John McKinlay and none other than Fredrick Walker, ex-commandant of Native Police in the Dawson. (chapter 3 volume 1). True to form, Walker was the only one to clash with the natives killing twelve and wounding several more, but according to records they were unlikely to have been Kalkadoons as Walker passed north east of their territory.

The first few settlers managed to establish cordial relations with the warlike tribe. Edward Palmer turned cattle up the Cloncurry River after finding the Gulf country grazing not to his liking. He established *Canobie* Station in 1864 on the fringe of Kalkadoon territory. Palmer was a learned man, an author, poet and keen amateur botanist. He represented Burke district on the Queensland Legislative Assembly until

1893 and was responsible for obtaining grants to establish schools and hospitals in Normanton, Cloncurry and Hughenden.

Palmer records no difficulties with the Kalkadoons. He gave them access to his holdings at all times and was openly critical of Native Police policies. On the other hand *Canobie* contained no areas sacred to the tribe as it was on the outer rim of their country.

Ernest Henry arrived as a prospector in 1866 after travelling through the country the year before with a mob of horses. He established relations with a party of Kalkadoons who led him to a large copper deposit in return for some clothes and a tomahawk. Henry established four mines in the area and persuaded Kalkadoon men, women and children to work one of them by supplying food and tools.

W. and T. Brown took up a small station they named *Bridgewater* in 1874. They too had no problems with the tribesmen but abandoned the lease shortly after due to impermanence of the water supply.

The rot began to set in when one of Queensland's more colourful pioneer figures Alexander Kennedy moved in to establish *Buckingham Downs* in 1877. By 1881 Kennedy had disposed of *Buckingham Downs* and in partnership with some others including Roger Sheaffe who came into the area with Ernest Henry, he obtained a controlling interest in the amalgamated properties of *Devoncourt, Bushy Park* and *Carlton Hills* by 1887 [30]. He ruled with an iron hand over 4,800 square kilometres of country carrying 60,000 head of cattle.

Alexander Kennedy had emigrated to Queensland from Scotland in 1861 at twenty-three years of age. His first job was at *Rio* Station on the Dawson then he moved to *Wealwandangie* on the Comet River where he worked for ten years becoming sheep manager of the property. No doubt the state of war on the Dawson and Comet had a bearing on Kennedy's attitude to the Blacks in the wake of *Hornetbank* and also *Cullin-La-Ringo* on the adjacent Nogoa watershed.

The Kalkadoons opened the batting around Christmas time in 1878 when a settler named Molvo and three employees were speared after they set up camp on a favourite permanent waterhole of the tribe on Sulieman Creek.

Sub Inspector Eglington and his troop were summoned from Boulia. they were met at the scene of the killings by Alexander Kennedy and a strong party of whites from nearby stations. They pursued the blacks into the nearby ranges where the Kalkadoon warriors contemptuously challenged the whites to fight. Not being all that well acquainted with firearms and their uses, scores of Aborigines were killed but none of the white party were left with the impression they had broken the resistance of the tribe.

For the next five years the law of gun and spear held sway in Kalkadoon country. Settlers were afraid to venture out unless in strong, heavily armed parties. The whites, male and female, carried at least one revolver and rifle at all times. Cattle were speared so continuously they became almost impossible to control and would rush (stampede) at the sight of a black person.

The natives proved themselves wily tacticians when they divided forces and attacked stations whose men were chasing decoy parties of Kalkadoons. The arrogant warriors took great delight in feasting on station cattle in full view and just out of range of the rifles of besieged womenfolk.

Alexander Kennedy travelled to Brisbane to seek more police protection for the beleaguered district resulting in the posting of sub-Inspector Marcus de la Poer Beresford and a division of Native Police troopers to Cloncurry. Beresford, an intense young man of aristocratic birth, took his duties seriously. He kept his troopers patrolling from one end of his district to the other attempting to punish the haughty tribesmen who jeered from the hillsides challenging them to come into the ranges and fight.

The one finger salute had not become a universal gesture in those days but the Kalkadoon warriors had no difficulty in conveying their contempt to Beresford and his troopers who were understandably reticent to take the fight to the Aborigines in their mountain stronghold.

On 24th January 1883 Beresford and his troop were camped on the headwaters of the Fullarton River in the McKinlay Ranges. Natives attacked but they managed to repulse them and the next day Beresford rounded the culprits up with an ease that would have made a more experienced frontier fighter a touch suspicious. The young commander decided to hold his captives overnight and felt quite secure with them yarded under guard in a gorge.

Like all good guerrilla forces, the Kalkadoons had caches of weapons stashed throughout their country. During the night, they armed themselves and fell on the black police killing Beresford and his men. One trooper got away and although carrying a spear head in his flesh, he made it to an outstation of *Farleigh* Run twenty miles away.

Eglington's Boulia troop arrived to deliver retribution but the tribesmen had vanished into their native hills. For a year after Beresford's death, the Kalkadoons had the upper hand in the war. Whites bristling with weapons were afraid to leave the immediate area of their homesteads. The natives speared or ran off all the police horses effectively immobilising the leaderless native troopers and Cloncurry barricaded itself expecting an attack on the town.

Blacks feasted on untended cattle from the stations and sent insulting messages to the whites, who cowered around their fortified dwellings. With the Cloncurry area as good as under Kalkadoon control, the Queensland Premier Sir Thomas McIlwraith, desperately searched for an officer who could bring order and restore the district to the settlers.

Sub-Inspector Frederick Charles Urquhart was a military college graduate who had served in the navy before arriving in Queensland where he joined the Native Police in 1882. In late 1883 Urquhart arrived in Cloncurry and zealously set about whipping the demoralised black troopers into shape. He obtained horses where he could and even mustered mobs running loose in the bush. According to old timers, he was a fearless rider who would throw a leg across the wildest brumby and show his men how to ride buckjumpers. Urquhart meant business and would do whatever it took to get his troop mobile.

The Kalkadoons regarded this flurry of police activity as hugely amusing. They sent a message with a town based Aborigine telling Urquhart to come out in the hills and they'd kill him as they had Beresford. The challenge is believed to have been sent by a warrior named Mahoni who may have been a Kalkadoon leader. Urquhart sent word back that the tribe would be wiped out unless they stopped fighting.

In mid 1884 the natives killed Alexander Kennedy's partner James White Powell while he camped on Mistake Creek. The warriors were jubilant. They had slain someone close to their sworn enemy Kennedy, and they feasted on his cattle and took photographs of White-Powell's family from the dead man's wallet to set up as spear targets.

Urquhart and his troop together with Kennedy and a man named A.F.Mossman of *White Hills* Station buried Powell, and Urquhart showed he had a poetic side by composing this piece of vengeful verse [31].

> *Grimly the troopers stood around*
> *that new made forest grave*
> *and to their eyes that fresh heap mound*
> *for vengeance seemed to crave.*
>
> *And one spoke out in deep stern tones*
> *and raised his hand on high.*
> *For every one of these poor bones*
> *a Kalkadoon shall die.*

They found the killers feasting on Powell's cattle and Urquhart, Kennedy and their followers struck the second decisive blow against the tribe. The police troop, assisted by settlers, never let up and for nine weeks after Powell's death, they waged unrelenting war on the Kalkadoons.

In September, a Chinese shepherd on *Granada*, one of the few stations in the area to run sheep, was speared and Hopkins the station owner, summonsed Urquhart and his men to the scene. Word went round the district like wildfire and parties of settlers primed their weapons and rode on the punitive raid they had determined would destroy the Kalkadoons.

About 600 warriors massed on a rocky hill that was soon to earn the name Battle Mountain. The warriors held the best defensive position but when the attacking forces tried a flanking movement, the Kalkadoons made an heroic blunder. Massing ranks, the warriors charged straight down at their attackers where they were shot to doll rags by withering volleys of rifle fire. Urquhart's diary makes no mention of the Kalkadoon charge but other accounts by competent researchers include it, so it should be taken seriously.

The battle of Battle Mountain that September day in 1884, finally destroyed the toughest fighting tribe in Australia twenty years after white settlement of their district. Two sayings exist in Australia today that celebrate bravery and were both inspired by players in frontier history: 'as game as Ned Kelly', and 'to fight like a Kalkadoon'.

Sub Inspector Urquhart went on to become head of C.I.B. in Brisbane but he was better at frontier fighting than forensics. His handling of the Gatton murders in 1898 was something of a shambles but Urquhart survived his critics becoming Brisbane's Chief Inspector in 1905, a position he held until 1917 when he became Police Commissioner.

In 1921 Frederick Urquhart was appointed Administrator of the Northern Territory holding the post until 1925 [32]. Urquhart died in 1932.

A Few Short Accounts of Aboriginal Resistance and White Reaction

To cover every known conflict between whites and Aborigines would fill several volumes. Conflict raged across the expanding frontier of Australia for one hundred and thirty years and small scattered incidents occurred even after the last notable clash in 1928.

The Twenty-Nine Year War

The Thungatti people of the Falls country in the upper watersheds of the Hastings, McLeay and Manning Rivers in north eastern New South Wales fought a war that lasted twenty nine years. Twenty retaliatory raids by settlers for white deaths and stock spearing between 1828 and 1853, produced 1,000 Aboriginal deaths accounting for one third of the Thungatti Nation and their associated sub groups[33].

When we consider the total number of Australians killed in the Boer, Korean and Vietnam Wars was around the same one thousand lives, it makes Mr. Justice Blackburn's ruling in the Northern Territory Supreme Court Land Rights Case of 27th April 1971 look a bit suspect. Blackburn's grounds for refusing Land Rights were that Australia was not a ceded or conquered Territory but was settled peacefully. Try telling that to what is left of the Thungutti.

Milbong and Dundalli – The Scourges of Early Brisbane

Brisbane's first armed robber Milbong Jemmy lost an eye and was hideously scarred from a fall into a fire in his youth. His name came from Milt (eye) Bong (gone) Jemmy (a name bestowed on Aborigines by whites dozens of times in early Australia). Milbong held the early settlement at Moreton Bay in terror during the 1830s and 1840s. 'Bumble' Gilligan the Moreton Bay Flagellator administered 50 lashes to the one-eyed Aborigine for robbing a miller at knifepoint. The punishment seemed to have little effect as Milbong's depredations continued. Throughout history beating never seems to have inspired a sense of admiration for the flogger in the floggee and Millbong Jemmy was no exception.

Jemmy was finally shot by two sawyers at Bulimba, now a Brisbane suburb and his head cut off for public exhibition. Milbong left an able apprentice, however, in Dundalli who took over as Moreton Bay's resident terrorist until he was captured in May 1854 [34].

Dundalli was such an important captive the Brisbane Courts imported Green, a hangman from Sydney, to do the honours. The highly regarded executioner managed to botch the job thoroughly. Dundalli swung choking at the end of the rope and Green was forced to complete his task by leaping on the unfortunate man's body to add his weight and finally break Dundalli's neck.

Warriors of the Lockyer

When white settlers moved onto Queensland's Darling Downs, 'Moppy' of the Yuggera who inhabited the Lockyer region to the west of Moreton Bay, was one of the area's principal chiefs. Moppy became a tribal 'Blood' brother to one of the squatters, John 'Tinker' Campbell, a native of Westbrook Maine in the U.S.A. The American and the Aborigine had huge respect for each other and 'Tinker' Campbell is reported to have claimed Moppy could raise 1,200 fighting men.

With the number of troops at 'Moppy's' disposal, if war had broken out in the Moreton Bay region, the whites may well have been driven out, but only scattered spearings of shepherds and stock occurred until 'Moppy's' son Multuggera, took over.

In August 1843 Multuggera formerly declared war on the whites as he was bound to do by tribal custom. His father's friend 'Tinker' Campbell was the obvious choice to make his declaration to. Campbell had gone into squatting as an early 'Aussie Battler' living by his wits and short of money unlike most of the 'Pure Merinos' who took up the Darling Downs, and his venture failed. Multuggera sought him out at a boiling down works he had started at Kangaroo Point in Brisbane, to declare his uprising in the traditional manner.

It was a naïve tactical move against whites but Aborigines conducted warfare in a ritualised way that satisfied honour and avoided unnecessary blood being shed. Multuggera declared his intention to spear all the military commanders and starve the newcomers out by stopping drays from travelling with supplies. His war opened with what has become known as the Battle of One Tree Hill near Helidon on 12th September 1843 where he put bullock drivers and armed guards of a supply convoy, to rout very convincingly. Like most white/indigenous conflicts in Australia and America, a few battles were won by natives, but the inexorable advance of the whites was unstoppable.

Multuggera's war became a series of skirmishes and eventually the Yuggera and their allies, the Limestone (Ipswich) Blacks who Multuggera recruited to the cause, went the way of all Aboriginal tribes becoming fringe dwellers of white society. Multuggera himself did not live to see the degradation of his people. He died in battle when he and twenty followers attacked John Coutt's *Rosewood* Station on 29th August 1846 [35].

Zac Skyring's Saviours

Despite the fact that a savage race war took place as their lands were overrun, Australian Aborigines proved many times they were a charitable, civilized people. This small incident demonstrates that.

When Gympie pioneer Zachariah Daniel Sparkes Skyring was a lad of thirteen in 1872, his father sent him to Brisbane to collect twelve cows and a bull and bring them back in company with a mob of cattle in charge of two drovers. Not far north of the city, the drovers stopped at a shanty and in the course of the resulting grog session managed to fall foul of the local law. With his travelling companions under lock and key, young Zac decided the only thing to do was to cut his charges out from the mob and head home alone.

While going through rugged country north of Nambour, something went wrong and the cattle scattered in the bush. Zac rushed around like a madman, but his charges had melted into the thick coastal forest vanishing like wraiths. The boy sat down on a log and began to cry at the thought of having to face his father with the news his cattle were lost. Suddenly, he became aware of Aboriginal voices in the surrounding scrub and having had many indigenous playmates as a child, Zac had a command of the local Gubbie-Gubbie language.

When he explained how he had lost the cattle and failed his father, the natives who placed great value on the honour of father and son in each other's eyes, slipped away in the forest and in a very short time had located the thirteen missing bovines. With his Aboriginal helpers, Zac made short work of the rest of the trip but on reaching the outskirts of Gympie, the Gubbie-Gubbie would go no further.

Gympie was a rip-roaring, gold mining town with the usual frontier attitude to Aborigines in those days, and Zac's saviours ran a very real risk of being shot particularly if seen driving cattle [36].

Conniston

Conniston Station to the northwest of Alice Springs in the Northern Territory was the scene of the last major Police action against Aborigines living in a tribal state in Australia. Mounted Constable William George Murray, a Gallipoli veteran stationed at Barrow Creek Police Station, drove out to *Conniston* on the Lander River with a tracker named Paddy, to investigate cattle spearing complaints by the station's owner Randall Stafford on 11th August 1928.

While Murray was there, word came to the station that Fred Brooks a dingo trapper, had been murdered by members of the Walbiri tribe at a soak in a dry river bed about twenty kilometres from the homestead. Murray drove to a telephone to report to Mr. Cawood, the Government resident for the area.

Cawood authorised Murray to deal with the matter and the policeman drove back to *Conniston* to raise a party and search for the murderers [37]. The next day Murray sent three men to verify the report and they found the body which had been hastily hidden in a large rabbit warren and re-buried it. Murray organised a party of seven men, Randall Stafford *Conniston's* owner, Paddy the tracker, a station Aborigine named Major, a young native called Dodger who had brought word of the killing, a half-caste named Alex Wilson and two whites Jack Saxby and Billy Briscoe.

The men left on horseback and finding the Walbiri camp, they lifted their horses into a run and tore right into the middle of the Aborigines. Naturally the natives were disturbed by the Hollywood-style entrance. Pandemonium reigned for a few moments with everybody yelling and screaming. Brisco suddenly began shooting. His mate Saxby opened up too and George Murray the policeman joined in. Three men and a woman were killed and another woman fatally wounded.

Randall Stafford was appalled and frightened about his own involvement. He returned to *Conniston* wanting nothing more to do with the business.

In the next week, Murray and his men ranged through the Lander River country indiscriminately killing Aborigines. The official tally was seventeen deaths but Murray is reported to have bragged the total was sixty or seventy [38]. This was verified when Aboriginal evidence of the crimes was finally taken in 1982 by a Walbiri Land Rights hearing [39].

George Murray was hailed as a hero in the outback where feeling was still strongly against Aborigines. A few months later in early 1929, the *Adelaide Register – News Pictorial* would describe him as 'the hero of Central Australia, the Policeman of fiction who rides alone and always gets his man.' The Walbiri knew Murray got his man alright, and woman and child too, for that matter.

Murray was involved in more Aboriginal deaths to the north of *Conniston* later on in 1928 and Athol McGregor, a lay Missionary campaigned for and eventually forced a government inquiry that determined at least thirty one deaths. The findings conveniently resolved that the shootings were justified [40].

Aborigines on Horseback

American settlers and soldiers faced the deadliest light cavalry forces since the Mongols during the western Indian Wars. Once Indians became mounted, the ease with which they could hunt buffalo ensured their food supply and they developed quickly into a formidable warrior society.

Considering how competently Aborigines embraced the horse once conflict was over and they settled on stations, their natural riding ability is obvious. Horse and gun placed Native Police troopers far above their tribal counterparts in battle, so white colonists should have been very glad that early Dutch and Portuguese navigators did not bring horses to *Terra Australis* as Spaniards did to the Americas.

The rapid spread of feral brumbies that occurred later, proved the majority of the country could support horses very well. Before the American plains Indians obtained the horse in the sixteenth century, they lived a very similar existence to Australia's tribes. The odd tribe like the Paiutes or Digger Indians of Utah, who did not adopt riding until they learned it from whites, show a resemblance to Aborigines in a tribal situation.

The question is purely hypothetical, but how formidable a foe would Kalkadoons on horseback have been? There is a fair chance if that had happened Alexander Kennedy

Photos of Australian Aborigines (above) and Paiute Indians (right) who did not become a horse culture.

Aborigines photo: Courtesy of A.M.Duncan-Kemp

Paiute Indians photo: Every effort has been made to trace ownership but without success. The source is acknowledged as Chancellor Press.

would not have stopped running east until he hit the water somewhere around Bowen.

American Conflict

Early Disputes

Some of the worst massacres of Native Americans occurred in the eastern states in the days of early settlement. The English colonists of Virginia who moved into the lands of the Powhatans, reduced the native population from 8,000 to 1,000 in the first few years of settlement [41].

The Dutch who bought Manhattan Island for a few fish hooks and glass beads, fell on two villages in the dead of night and killed every inhabitant. America invented 'Manifest Destiny' - a politically expedient catch phrase to excuse taking whatever they wanted, and although the pious New Englanders of Massachusetts who were kept alive by Indian charity in their first year, argued against the principles contained in the phrase, they also managed to wipe out every tribe in the area they claimed.

Andrew Jackson, as President in 1829, established the 'Permanent Indian Frontier' from the Mississippi River to the 95th Meridian on the Canadian border. The 'Permanent Frontier' proved amazingly impermanent as settlers rushed west with or without sanction. The situation was much the same as the Australian experience of the same year when the Government applied its Limit of Settlement ruling. The plan to confine settlement to nineteen hastily formed counties following the movement of land takers into the country west of the Blue Mountains in New South Wales, was no more successful than the American attempt to contain their settlers [42].

The Battle of Beechers Island

In the winter of 1867 – 68 Southern Cheyenne and Arapaho tribes were camped near Fort Larned below the Arkansas River in Kansas. Most of the chiefs had signed a treaty on 14th October 1865 in which they agreed to live in peace south of the Arkansas.

The Cheyenne had been given a graphic illustration of army tactics four years previously when at dawn on 29th November 1864 former Baptist Minister Colonel John M.Chivington and 1,000 troops of the 3rd Colorado Volunteers, fell on the sleeping camp of Black Kettle's band at Sand Creek in Colorado Territory [43].

The five hundred inhabitants of the lodges were surprised to say the least. They had been told a peace treaty was in effect and had surrendered their weapons at Fort Lyons some days before.

Chivington and his men swept down on the camp at dawn killing 150 men, women and children and the 'heroic' soldiers later paraded through Denver waving severed scalps and genitals of their victims.

The Sand Creek Massacre and other engagements had broken the power of the local tribes and in the face of the railroad and the buffalo hunters it brought in its wake, the Indians had little option but to agree to the treaty of 1865.

Washington was always better at making treaties than adhering to them. Food and clothing were not forthcoming and soon bands of angry young men were slipping back across the Arkansas to raid and hunt. By August 1868 about 300 Cheyenne warriors and their families were gathered along the Arikaree Fork of the Republican River under Chiefs Tall Bull, White Horse and Roman Nose, as well as a few Arapaho and Sioux.

The Kansas/Pacific Railroad was halted at Sheridan Station west of Hays City due to finance problems. Soldiers, track workers and buffalo hunters were hanging around the rail head with time on their hands and like people with idle hands, they soon dreamed up some devil's work.

Major George Forsyth was an aide to General Phil Sheridan fresh from West Point who had never fought Indians before. Spoiling for some action, he and Lieutenant Fred Beecher, an army surgeon, raised a civilian company of fifty buffalo hunters, scouts and others who happened to be hanging around Fort Wallace at the time, for a little diversion hunting 'pesky redskins' [44].

Forsyth had no trouble gaining the blessing of General Sheridan who favoured a policy of extermination towards all Native Americans. The General even dignified the rabble with an official title – 'Forsyth's Scouts'.

The expedition made its way north and set up camp on the dry Arickaree Fork near a small sandy island. Two Scouts were sent out to reconnoitre – Sharp Grover an Army Scout, and a young man named Jack Stillwell, but they failed to discover the Cheyenne encampment nineteen kilometres downriver.

At dawn on 17th September, an ominous rumble of hooves sounded as Forsyth, Beecher and their men stood around the cooking fires. They looked up in alarm as a small party of whooping Indians swept over a low ridge and tried to run off their horse herd. It was a few headstrong young braves and although they got some horses, it alerted the whites and they quickly shifted to the small sandy island and dug in for the defence.

Forsyth and his men soon realised they were in for more sport than they had bargained for when the main body of Cheyenne charged the island.

The Indian forces split in their classic horseback attacking formation, and began circling Forsyth's Scouts who lay behind packs in hastily dug breastworks returning the Indians' fire. The whites' hearts leapt in sudden expectation when the strident notes of a bugle rang over the battlefield but their hopes were as suddenly dashed when they saw it was blown by a grinning Negro in full war paint riding with the Cheyenne.

One Cheyenne warrior named Wolf Belly made two individual charges right through the Scouts' defensive ring and incredibly was not touched by a bullet. Forsyth and his men were not nearly as cocky as they started out. One, who had plenty to say before the attack, dug a hole which he tried to crawl into, giving a fair imitation of an ostrich.

The Scouts' horses panicked and those that didn't fall to the arrows and guns of the Indians, broke and ran. Lieutenant Beecher was dead. Forsyth was wounded with a broken leg and the party was in desperate straits when an imposing Indian appeared on the overlooking ridge. This was Roman Nose, a War Chief of huge standing among the Southern Cheyenne. Like Pemulaway of Australia's Eora Aborigines, his people believed him to be invincible to the white man's bullets. Roman Nose carried great medicine. He was a holy man regarded in almost immortal terms among his followers, but this day his armour had a chink in it.

At a feast put on by the Sioux a few days before, food Roman Nose had eaten was prepared by a Sioux woman who used an iron fork not knowing it was against the War Chief's medicine. Cheyenne belief meant the touch of iron would void his magic power against the whites' weapons, but Roman Nose painted for war anyway, knowing he had no time for a purification ceremony.

Like Aborigines who would die if they believed the bone was pointed at them, Roman Nose knew in his heart he was doomed, but he led the next charge anyway. The front ranks went down before the fire power of the whites and still Roman Nose drove his war horse forward. Right to a scrubby fringe of willows almost under the guns he went, until a cross fire caught him through the hips and he fell mortally wounded. Indian women crept up under cover of darkness to carry their Chief away and Roman Nose died during the night.

The loss of their leader took all heart for fighting away from the Cheyenne, but still they kept Forsyth's Scouts pinned down for eight days. The white men were forced to eat the putrid meat of their dead horses and dig for water [45]. Had it not been for a relief column of soldiers who arrived on the eighth day, Forsyth's foolhardy crew would certainly have perished. The Scouts suffered twenty-three casualties and the Cheyenne about thirty and the low sandy islet in the Arikaree fork of the Republican took the name of the foolhardy Lieutenant who died there on Beechers Island.

Custer's Last Stand

The best-known battle between whites and indigenous people in America and probably the world, is the resounding defeat of Lieutenant Colonel George Armstrong Custer and a detachment of 7th Cavalry by a combined force of Sioux, Cheyenne and Arapaho Indians at the Little Big Horn River, Montana Territory on 25th June 1876 [46].

FIRST ACCOUNT OF THE CUSTER MASSACRE

TRIBUNE EXTRA.

Price 25 Cents.   BISMARCK, D. T., JULY 6, 1876.

MASSACRED

GEN. CUSTER AND 261 MEN THE VICTIMS.

NO OFFICER OR MAN OF 5 COMPANIES LEFT TO TELL THE TALE.

3 Days Desperate Fighting by Maj. Reno and the Remainder of the Seventh.

Full Details of the Battle.

LIST OF KILLED AND WOUNDED.

THE BISMARCK TRIBUNE'S SPECIAL CORRESPONDENT SLAIN.

Squaws Mutilate and Rob the Dead

Victims Captured Alive Tortured in a Most Fiendish Manner.

What Will Congress Do About It?

Shall This Be the Beginning of the End?

Headlines from the Bismark *Tribune Extra* (South Dakota) following the defeat of Lt.Col. George Armstrong Custer

Custer, a native of New Rumley, Ohio born in 1839, always dreamed of a military career. He entered the West Point U.S.Military Academy in 1857 and graduated bottom of his class in 1861. He achieved the glory he craved in the Civil War rising to the rank of general in the Union Army, by the age of twenty three and was reduced in rank to Captain after the South's surrender.

Custer went west to fight Indians where he received promotions to Lt.Colonel and his first notable encounter was the Battle of the Washita on 27th November 1868 in which he reported one hundred warriors killed. Major General Phillip 'Little Phil' Sheridan praised Custer's gallantry and bravery during the engagement but Sheridan's favourite saying was "The only good Indian is a dead Indian." The tally of dead Southern Cheyenne at the Washita was one hundred and three, but only eleven were warriors. The rest were old people, women and children [47].

George Armstrong Custer was a frontier dandy with long, curling, blonde locks and a penchant for fancy non-uniform garments like fringed buckskin jackets. He was also a glory hound with political aspirations, who believed a successful campaign against the Sioux, Cheyenne and Arapaho would be his ticket to the White House.

Custer ordered his regiment into three columns at Little Big Horn. One under Major Marcus A.Reno; one led by Captain Frederick W.Benteen; and one led by himself. It appears he was too eager to engage the enemy and became separated from the other columns by around eight kilometres. He was surrounded on a hillside and his command wiped out to the last man. So instead of becoming President, Custer just became deceased.

Custer's Conquerors

The Native American generals of the army that defeated Custer at Little Big Horn were all fighting men of the first water. There were many leaders among the Sioux, Cheyenne and Arapaho forces who opposed the armies commanded by Generals Alfred H.Terry and George Crook during the campaign against the northern tribes in Montana and Dakota [48]. Indian Chiefs could not command their men. Following one on to the field of battle was purely voluntary for all warriors who wished to go. To attract fighters to a cause, Native American leaders needed to be popular and charismatic.

Naturally many Chiefs rose but the most recognised leaders of the various people involved, were Red Cloud and Spotted Tail of the Teton Sioux, a nation comprised of the major sub-groups Oglala, Brule, Hunkpapa and Minneconjou. The Northern Cheyenne were represented by Dull Knife, and the Arapaho by Little Raven.

Two Chiefs, who would become legends for their fight against invasion of their sacred Black Hills, were a Shaman and War Chief of the Hunkpapas, Tatunka Yotanka or Sitting Bull, and the Oglala Tashunke Witko or Crazy Horse.

Sitting Bull

Tatanka Yotanka means sitting buffalo bull if its full translation is used, but the world has come to know the charismatic Indian leader of the Hunkpapa Sioux as Sitting Bull. The Shaman and War Chief was born in the Grand Rivers region of South Dakota in 1834 and was known in his childhood as Jumping Badger [49]. His first notable contact with white soldiers was a raid against Fort Buford in 1866 and he rode the war path with his followers from 1869 to 1876 almost continuously. When not raiding the whites, Sitting Bull led forays against traditional Sioux enemies, the Crow and Shoshone tribes. Sitting Bull liked to fight.

It was Sitting Bull's refusal to take up reservation life that became the excuse for General 'Little Phil' Sheridan ordering Generals Crook and Terry into action against hostile Indians in the Tongue, Powder, Rosebud and Bighorn Rivers region of Montana and Dakota Territory [50]. Sitting Bull was the War Chief of the Hunkapapa in that historic clash against the 7th Cavalry and led the fight against Major Reno's Division.

It is due to testimony from Sitting Bull that one of the popular myths about Custer's final battle was questioned. When interviewed in Canada a year after the battle, he told how the flamboyant Lieutenant Colonel was not immediately noticeable amongst the dead. The habit of frontiersmen wearing their hair long is believed to have come about as an act of bravado against Indians. A scalp with long flowing locks was a fine trophy of war and despite the fact Native Americans learned to scalp from early white bounty hunters who took scalps to collect payment, the practice became widespread among them.

George Armstrong Custer grew his golden curls when he went west to fight Indians supposedly as a challenge to enemies to 'take his hair' if they could. On the eve of Little Big Horn, he may not have been feeling as tough as he liked to make out because Sitting Bull testified that 'Yellow Hair's' body sported a brand new haircut [51].

Following the battle, Sitting Bull escaped into Canada where he lived until 1881 when he was persuaded to return under a promise of amnesty. It is ironic that he surrendered at Fort Buford, the scene of his first engagement with the U.S. Army. After a period of confinement at Fort Randall, the Chief lived in a cabin near the Grand River, part of the Standing Rock Reservation in Dakota. He appeared as part of Buffalo Bill Cody's Wild West Show in 1885 touring cities in the United States and Canada.

Sitting Bull's colleague in the show, sharpshooter Annie Oakley, told of him giving his money away to homeless people and of being appalled by the poverty in the white man's cities. He was an immensely humane person who continually warned his people against rejecting their culture in favour of that offered by the whites. An Australian who met him in the 1880s described him as a 'kind but sad man' [52].

Sitting Bull gave a fifteen-city lecture tour where he spoke through an interpreter, met President Grover Cleveland and ex-President Ulysses S.Grant, and received fan and hate mail from around the world [53].

Indian Agent, Major James McLaughlin at Standing Rock, did not like Sitting Bull at all and the Hunkapa Chief returned his sentiments. In 1888 Sitting Bull advised the tribe to refuse an offer to purchase their land and most of the wilder elements of the defeated Sioux camped close to his cabin.

The Indians' last hope of a return to their traditional lifestyle came in the Ghost Dance Religion. A Paiute Indian mystic named Wovoka claimed he had a vision from the Christian God during a solar eclipse of the sun on 1st January 1889. Wovoka's vision showed the whites disappearing from the earth and all native people living or dead being re-united to live in peace and harmony. He claimed, to bring this about, Indians must adopt his doctrine and practice the sacred 'Ghost Dance'.

The Ghost Dance Religion swept through Native American Reservations in 1889 and 1890, and Sitting Bull invited Kicking Bear of the Minneconjou Sioux to organise the first dance at Standing Rock. An interesting Australian parallel to the Ghost Dance is told in Aboriginal rock art in caves among the Conglomerate Range on Queensland's Cape York Peninsula. The paintings tell of a being with supernatural powers, known as the 'Quinkan', who would rise and wipe out the white invaders [54].

Alarmed by the Ghost Dancing, which he saw as the prelude to a revolt, James McLaughlin ordered Sitting Bull's arrest. When members of the Indian Police Force responsible for law and order on the reservation attempted the arrest, Sitting Bull's people tried to rescue him. In the ensuing struggle, Sergeants Red Tomahawk and Bullhead of the Indian Police shot and killed the Hunkpapa Chief on 15th December 1890. Sitting Bull's son, Crowfoot, and six of his other followers died in the fight as well as seven members of the Indian Police including Bullhead.

Crazy Horse

The man who would become one of the best known figures in the Indians' struggle against white invasion was born on Rapid Creek at the eastern end of Paha-Sapa, the sacred Black Hills of the Sioux, sometime between 1840 and 1845. His boyhood name was Curly and in the Indian custom, he went to live with his uncle High Back Bull or Hump of the Brule [55]. He joined his first war party at age sixteen in a tribal raid against the Gros Ventre Indians of the northern Montana/Canadian border country.

Crazy Horse was a mystic who spent much time in prayer and meditation. His visions were legendary among his people and it is believed to be one such vision in which he saw his horse dancing as if it were insane, that was responsible for the warrior name he took after his initiation to manhood [56].

His first memorable coup against soldiers took place during the siege against Fort Phil Kearny on the Bozeman Trail, Montana on 21st December 1866 when he led a decoy party that lured a brash young officer, Captain William J. Fetterman, and his command of around eighty men, to their deaths [57]. Fetterman had bragged that with eighty men he could ride through the whole Sioux Nation but he had just left sight of the Fort, when Crazy Horse's decoy party led him and his men to their deaths.

In the Battle of the Rosebud on 17th June 1876 he led 1,000 warriors against 1,100 soldiers under General George Crook's command and handed Crook the worst defeat he had suffered in both the Civil War and Apache Wars in the southwest. Eight days after Crook's humiliation at the Rosebud, Lieutenant Colonel George Armstrong Custer and his men appeared on a rocky bluff overlooking the huge camp of Sioux and Cheyenne that stretched along the Little Big Horn River.

Crazy Horse and his followers swarmed out of camp to cut Custer off from the ford he was making for and surround him in the most famous Indian battle of all – Custer's Last Stand.

In July of 1877, Crazy Horse was persuaded to come in to Camp Robinson in Nebraska to surrender with several thousand of his followers from the Sioux Nation. He was betrayed by his own people Red Cloud and Spotted Tail whom the Army had designated as the principal Sioux Chiefs and put each in charge of a separate Agency or Reservation. The two Chiefs became jealous of Crazy Horse's popularity.

On 4th September, Crazy Horse left Fort Robinson to take his wife, Black Shawl, who was very ill with tuberculosis, to her parents' people at the Spotted Tail Agency. His enemies claimed he was fleeing custody and Captain Lea at the Agency urged him to return to Camp Robinson to explain his actions.

He walked into the Camp Robinson administration building on 5th September flanked by Captain James Kennington and one of the Indian Agency policemen, Little Big Man, with his cousin Touch the Cloud in front of him. Touch the Cloud became suspicious and cried out "Cousin! They will put you in prison."

Little Big Man and the Captain seized Crazy Horse's arms and an un-named soldier drove a bayonet into him from behind [58]. Crazy Horse died from his wound that night. His father sang the death song over him and he was spirited away to a secret grave in the Dakota Bad Lands. The location of the grave is a mystery to this day.

The Avengers

America had a few whites who pursued personal vendettas against Indians in a similar manner to Australia's William Fraser. One shadowy figure in folklore was the inspiration for the film *Jeremiah Johnson* starring Robert Redford. John Johnston was a free trapper around the end of the fabled Mountain Man era who returned from a winter trapping expedition to find his Indian wife and child murdered.

After reading tracks around the cabin, Johnston deduced the killers to be members of the Absaroka or Crow tribe. In a one-man war, John Johnston earned the name of 'The Crow Killer' by riding through the middle of the Absaroka Nation killing warriors right and left. He was also known as 'Liver Eating Johnston' because he is supposed to have once eaten the liver

The grave of John Johnston 'The Crow Killer', Cody, Wyoming, USA

Photo courtesy John Osbourne collection

of a noted warrior he slew in a hard fight. Legend has him killing 300 Crow warriors but this is almost certainly a wild exaggeration.

'Liver Eating' Johnston was a wild man right to the end. The story is told that as an old man he once purchased the services of a lady of the evening in a Montana country town[59]. It seems when the ill matched pair tumbled into the lady's bed, the mountain man, suffering from a mixture of old age and previously consumed liquor, failed to rise to the occasion.

Johnston was soon snoring in oblivion and the girl returned to the bar downstairs to regale its patrons with the tale of the 'Liver Eater's' shortcomings. The old man awoke and annoyed by the absence of his paramour, he took a dress that hung behind the door and using a broom handle, stirred it thoroughly around in the unemptied chamber pot that resided beneath the bed.

Johnston returned to the bar to be greeted by hoots and catcalls from all present particularly the girl. The old man waited until the racket died down then addressed his former, bed mate. "Don't worry honey" he told her. "You got better screwed than you know".

A reporter from the local paper heard the story and printed a tactfully worded version under the heading "A Scout's Revenge: Stirring Times in the Old West".

An early inhabitant of Texas named Jeff Turner came from Kentucky to settle on the Guadalupe River. He returned from hunting one day to find Indians scalping his dead wife and three sons. Turner fell on the killers taking four lives and from then on he lived for nothing but killing Indians.

Turner became known as 'The Indian Hater'. Texas Ranger Captain 'Bigfoot' Wallace described him as a tall, spare built frontiersman with matted hair dressed in buckskins and a coonskin cap whose eyes glittered, flashed and danced around in his head. He rode a vicious looking, raw boned horse called 'Pepper-Pot' and carried a muzzle-loading Kentucky rifle ready for instant use.

Jeff Turner is supposed to have taken forty-six Kiowa and Comanche scalps in the days before Texas became a State in 1845, and he eventually fell foul of the Indians he relentlessly pursued.

According to Bigfoot Wallace, Indians came on Turner asleep and took him captive. The Indian Hater was staked out and the chief cut his heart out and gave it to his pregnant wife to eat in the belief Turner's reckless bravery would transfer to the unborn child[60].

In the early days of Kentucky settlement, a teamster called Simon Girty turned against his own race and threw his lot in with three eastern tribes. Girty had been captured by Indians when a fourteen year old, around 1770. He returned to white society but continued to move between both races[61]. Girty helped command combined Mingo, Delaware and Wyandotte tribes as an advisor to their Chiefs when they handed a crushing defeat to General Arthur St.Clair who led a punitive expedition authorised by President George Washington in 1791.

Washington despatched St.Claire against 'certain banditti of Indians from the north west side of the Ohio' and when Girty led hostile Indians through the Ohio River Valley, he is reputed to have casually watched a former friend burn at the stake[62]. Simon Girty fled to Canada when the tribes were eventually subdued, where he died crippled by rheumatism and insanity in 1818.

Geronimo and the Last Wild Indians

On 17th February 1909 the legendary Chiricahua Apache War Chief, Geronimo, known to his people as Goyathlay, died at Fort Sill, Oklahoma. Geronimo was one of the last hostile Indians. He was brought in to the Chiracahua Reservation but when it was abolished in 1876 and the Apache transferred to San Carlos in New Mexico, Geronimo fled into Mexico with a band of followers.

He was captured and returned to the new Reservation but escaped again in 1881, to begin raiding with a group of other Apache Outlaws. He surrendered to General George Crook's Scout Tom Horn (chapter 4 volume 1) and came in again. In 1885 he ran again and made war on the whites for a year until Crook once again arranged his surrender. At the last moment, Geronimo fled and was not re-captured until 1886 when Crook's successor General Nelson A.Miles brought him in.

Geronimo and his fellow Chiefs were transported to Fort Marion in Florida as prisoners of war. Consistent lobbying by his old adversary General Crook and Indian Agent John Clum who fought long and hard for Indian rights, succeeded in having them brought back to Oklahoma. It was not their homeland, but far preferable to the fever-ridden Florida Coast[63].

There is a persistent rumour that Geronimo sent the remnant of his people to a hidden valley in Mexico's Sierra Madre Mountains while he allowed himself to be captured to throw General Miles' soldiers off their tracks. In 1913 Mexican Revolutionary, Pancho Villa, told General Hugh Scott of the United States Army, that he knew of Wild Apaches living in the Sierra Madre and in 1934 a group of young Apache warriors reputedly led by Geronimo's grandson, raided a Mexican village[64]. They were reported to have come out of the Sierra Madre.

WHITE AGAINST WHITE

So we must fly a rebel flag as others did before us
And we must sing a rebel song and join in rebel chorus.
We'll make the tyrants feel the sting of those that they would throttle.
They needn't say the fault is ours if blood should stain the wattle..

Henry Lawson

The Difference Between Two Continents

The United States of America was created by the War of Independence, which ended British Rule in 1781. Fifty-five years later, Texas gained independence from Mexico with the most famous engagement being the historic 'Alamo'. The fledgling Republic was again plunged into conflict from 1860 to 1866 during the Civil War. In fact many battles occurred between white Americans.

Early uprisings between rival fur trading companies were followed by bitter range wars that flared up between settlers and big pastoral companies, the opposing interests of cattle and sheep, and even a battle of religious origins. White Australians on the other hand, have only joined in pitched battle on Australian soil three times: at the Battle of Vinegar Hill in 1804; at Ballarat's famous Eureka Stockade in 1855; and at *Dagworth* Woolshed in Western Queensland in 1894.

Australia also saw a few race riots involving whites and other non-indigenous races.

There is no doubt America has seen more major battles and feuding factions than Australia, but human nature being what it is, man has attempted to resolve differences on the battlefield on both continents.

Conflict Down Under

The Battle of Vinegar Hill

To gain an understanding of events leading up to the Battle of Vinegar Hill we need to go back to the homeland of its most well known participants – the Irish. The depredations of Oliver Cromwell in 1649 followed up by Ireland's defeat by William of Orange in 1690, caused a hatred against the English that remains to this day by children of the Emerald Isle.

The rise of the British Empire was a bitter pill to the Irish, but they held to their language and religion and never let their dream of again achieving independence, die. Encouraged by England's defeat at the hands of America's new colonists who won their revolution in 1781, anti English rhetoric and plotting flared into open rebellion when the Irish made a tragic and unsuccessful attempt to reclaim their land.

Pikes and pitchforks were no defence against the explosive Howitzer shells of English artillery and on 21st June 1798 the Irish revolution was crushed in a bloodbath at Vinegar Hill outside Enniscorthy [1].

Remnants of the Irish rebels were shipped in chains to New South Wales. Among them were Phillip Cunningham and William Johnston from County Kerry. Cunningham and Johnston were true insurgents who saw possible death as a far superior alternative to certain bondage, and they began laying plans and enrolling supporters for an Australian Revolution.

Farms around the fledgling districts of Parramatta and the Hawkesbury River were their main recruiting areas and soon a motley crew of convicts from many races including English, were recruited to the cause.

The first strike was to be at Castle Hill were it was planned that labourers would overpower their employers when they saw a signal fire, then take Parramatta and advance on Sydney itself [2]. News of the uprising leaked out to Captain Edward Abbott, commander of a detachment of the New South Wales Corps stationed at Parramatta. Abbott's overseer, a man named Sloane, brought the news that the rebellion would take place on Sunday, 4th March 1804. Abbott took Sloane's statement and conveyed it by messenger to Government House in Sydney on Saturday 3rd.

The opening engagement at Castle Hill was a heady draught for the would-be rebels. Cunningham and Johnston put their plans into effect at 8 o'clock on Sunday evening. They took possession of the settlement and a building in the Government compound was fired as Cunningham marshalled his two hundred odd recruits and searched the area for arms.

To the delight of the ragtag army, a known and hated flagellator named Robert Duggan was found hiding under a bed and dragged to the flogging triangle for a bit of role reversal. Duggan had killed prisoners before by his brutal application of the lash and although the rebels didn't take his punishment that far, Duggan gained a very real insight into his victims' sufferings as the 'cat-o-nine-tails' laid his back open.

Recruits steadily filed in but without the rebels' knowledge one of the Castle Hill auxiliaries had escaped to Parramatta to alert the garrison there. A detachment sent by Cunningham to fire the house of John Macarthur (the founder of Australia's wool trade), and draw troops from the city centre, lost their way and the fire was never lit. A time of confusion ensued and it soon became obvious that Parramatta was awake and defended, so the plan was changed. The revolutionary forces decided to launch their attack on the Hawkesbury instead.

Despite a night without sleep, the rebels were still in high spirits. They took the track north-east to the Hawkesbury and as they marched, new recruits straggled in to join them. By the time they had reached a place named Second Ponds Creek near present day Windsor, their numbers were around 400.

Meanwhile Captain Abbott's messenger had reached Sydney and Governor Phillip Gidley King was in possession of the news that his Colony was under attack. King hardly had time to digest the information when he received another message that Castle Hill had fallen and armed insurgents were marching on Parramatta.

The Governor called out the English loyalist militia and sent word to Captain Woodriff of the warship '*Calcutta*' that was lying at anchor in Sydney Harbour, to send all the armed men he could. He appointed Lieutenant Governor William Paterson to take charge in Sydney and took horse for Parramatta stopping en route at Major George Johnston's property to enlist his support in bringing troops to the area.

King arrived in Parramatta at four in the morning and Major Johnston followed an hour later with what men he could muster. At Government House, King declared martial law throughout the districts of Parramatta, Castle Hill, Toongabbee, Prospect, Baulkham Hills, Seven Hills, Hawkesbury and Nepean. He and his men then stood ready to repulse the rebels who by now were encamped on a Hill near Second Pond's Creek that would soon be known as Australia's Vinegar Hill.

Johnston, whom King placed in charge of the troops, soon found out the rebels' intended destination from settlers who straggled in after fleeing before Phillip Cunningham and William Johnston's army. He marched his hastily assembled forces towards the Hawkesbury and soon came upon the rebels formed up in four columns on the slopes of Vinegar Hill.

Major George Johnston was an experienced soldier. He had seen action in the American War of Independence and the East Indies before joining a Marine detachment in New South Wales, and was claimed to be the first member of the First Fleet to set foot on Australian soil.

The morning was well advanced when Major Johnston lined his red coats up to meet the rebels. Johnston attempted to negotiate a surrender under a flag of truce, by sending an emissary named Trooper Handlesack to meet with the rebels. Handlesack returned with an emphatic "No!" The Major then sent a Catholic Priest named Father John Dixon with the same message, which was greeted by howls of derision from the Irish Convicts. To their minds, no Catholic churchman should carry messages for the hated English. It has never been established whether Father Dixon was there by choice or under duress.

The Major then rode within gunshot under a flag of truce and challenged the rebel captains to come out and meet with him. Phillip Cunningham and the Major's namesake William Johnston met and parlayed with Johnston midway between the assembled armies and on the promise of bringing Father Dixon up again, Major

Johnston returned to his lines and came back with the priest and his Quartermaster, Laycock.

Talk again proved useless and when Phillip Cunningham gave his final reply by yelling "Death or Liberty!" The Major treacherously whipped out a pistol concealed in his clothing and held it to Cunningham's head. At the same moment, in a move that had obviously been planned, Laycock also produced a pistol and held it on William Johnston.

The men were dragged back to the English lines then Major Johnston ordered an immediate attack. The leaderless rebels wavered in the face of the red coats and although they managed to return fire, the soldiers poured another volley into their ranks and followed with a bayonet charge that left over twenty of the would-be revolutionaries dead or wounded. A further twenty-six prisoners were taken and Johnston's men pursued the routed rebels to Windsor where they lynched Phillip Cunningham who was badly wounded and possibly already dead by hanging him from a staircase. He had been taken prisoner in good health.

Three days later, William Johnston and nine other insurgents were sentenced to hang at a Court Martial presided over by Captain Edward Abbott. A further seven rebels were sentenced to 2,600 lashes between them, and another two Dennis Ryan and Bryan Riley who were wounded, were handed down a rather open ended sentence that specified they receive as many lashes as they would stand without dying.

Thirty more Irishmen were sent to the Penal Settlement at Coal River (Newcastle) and sentenced to total silence and work without rest except to eat and on Sundays. More went to the living hell of Norfolk Island.

A final act of retaliation was the total suppression of Catholicism in New South Wales for the next decade.

Major George Johnston went on to obtain further land, wealth and fortune in New South Wales and ironically, four years later, he became a rebel himself in Sydney's Rum Rebellion on Australia Day 1808.

Backed by John Macarthur, Johnston led the military coup that deposed Governor Bligh of '*Bounty*' fame, and seized power for six months. Macarthur's rebels were a lot better organised than the Vinegar Hill boys and they not only largely escaped retribution, but became rich in the process.

Colonial administration attempted to wipe the battle of Vinegar Hill from memory. It was only included in history books as a much played down Irish convict riot, and the scene of the historic battle was re-named Rouse Hill.

The Eureka Stockade

In August of 1851, the fabulous Ballarat Gold Field in Victoria was discovered and a plethora of fortune hunters from practically every race and creed soon descended on the new Victorian Gold Fields [3].

All the problems that beset gold rushes around the globe were present at Ballarat. Discrimination against Chinese who were by far the largest aggregation of foreigners; disputes over boundaries; claim jumping; and as choice a band of cutthroats, whores, con artists, sly-groggers and thieves as were ever assembled.

On the Ballarat field, however, the worst thieves and stand-over men in the miners' eyes, were Victorian Lieutenant Governor Charles Joseph La Trobe who imposed a thirty shillings per month licence fee on all diggers in 1851, and Goldfields Commissioner Doveton and his squad of police who enforced collection.

Thirty shillings per month was equal to the annual wage of a menial worker in those days and a terrible tax to impose on the struggling gold seekers. Miners sent a deputation to complain to the Commissioner, but Doveton laid down the law to them stating flatly "If they failed to pay the licence, he would damn soon make them." In the new year Governor La Trobe doubled the fee to £3 per month. It seems English administration had failed to learn from their mistakes in America where similar savage taxes had been one of the main causes of the revolution in that country, and Ballarat soon became a hotbed of potential rebellion.

The *Melbourne Argus* named La Trobe as a 'foppish whippersnapper' and Doveton and his troopers as 'mushroom puppets of a rotten government.' [4] Protest meetings were held in Bendigo, Ballarat and Melbourne and under pressure La Trobe did a backflip and reset the tax at thirty shillings.

The Governor had backed down as far as he was going to though, and probably as a salve to wounded pride, he instituted the practice of 'digger hunts' to ensure collection of his thirty bob a month. Police and soldiers swept through the fields on a regular basis demanding that licences be presented. Diggers unable to produce them were arrested, fined and in some cases, assaulted. The approach of troopers often resulted in a mad scatter as 'illegals' dived for cover wherever they could find it [5].

In August 1853, over 3,000 diggers gathered at Ballarat to hear the outcome of a petition presented to the Governor protesting the fees. They assembled under the flags of their various nationalities and a colourful sight it must have been with flags of all nations flying and the bearded diggers alternatively cheering and jeering the points put forward by spirited orators who harangued the crowd.

The antagonism shown by the miners threw a scare into Governor La Trobe and in November 1853 he proposed to the Legislative Council that licence fees be done away with and replaced by an export duty on gold. The Council comprised of large-scale landowners, had little regard for the problems of miners whom they regarded as a 'rabble', and La Trobe's proposal was rejected.

La Trobe was probably pleased to hand over the Governorship to Sir Charles Hotham in June 1854, but the change of leadership did nothing for the Ballarat diggers as Hotham's inclination was to deal with the situation by use of the military.

In August he visited the gold field and after saying encouraging enough things to the assembled men to warrant a cheer or two, he took his leave and the following month, Hotham ordered licence inspections to be performed fortnightly instead of monthly.

The zealousness shown by police in enforcing licence laws left them very little time for anything else and crime became rampant on the diggings. Many of the troopers were time-expired convicts and there is abundant evidence to suggest police corruption was in epic proportions on the Victorian gold field.

An hotel named The Eureka had an unsavoury reputation and when James Scobie was murdered there on 6th October 1854 his killer was acquitted by Police Magistrate Dewes. Dewes was known to owe money to J.F.Bently the owner of the Hotel who was witnessed wielding the shovel used to take Scobie's life. James Francis Bently was a time expired convict who had a knack for cultivating people in prominent positions. He had loaned Magistrate Dewes money to indulge in land speculation, as had several hotel proprietors on the unstated understanding that they might have trouble renewing their liquor licences if the request for money was refused.

Less than two weeks later, 5,000 diggers met to protest the event and form a committee to demand a retrial of Bently. Following the gathering a large mob of the more militant miners descended on the Hotel and burnt it to the ground. This incident is generally regarded as the catalyst for the revolt that was to follow. During the riot Sub Inspector Maurice Ximenes lent his horse to Bently who escaped leaving his hotel and pregnant wife behind [6].

Andrew McIntyre, Thomas Fletcher, Henry 'Yorkey' Westerby and five other ringleaders, were charged with arson and rioting. The flags flew again on Saturday 11th November as bands struck up at another open meeting that saw the formation of the Ballarat Reform League described by the *Ballarat Times* as 'the germ of Australian independence'.

Hotham refused to listen to a deputation requesting the release of the prisoners and when he ordered more troops from Melbourne, angry diggers met them on the road on 28th November. Captain H.C.Wise leader of the new detachment, refused to speak with the men whereupon they attacked the convoy overturning a wagon and killing several soldiers.

The troopers retreated to the military encampment and when further soldiers were sent to punish the rebels, they were fired upon from the diggings and pushed back to their encampment.

On the evening of that first armed clash between soldiers and miners, the Reform League met and an Irishman who was to become one of Australia's most lauded heroes, Peter Lalor, was democratically elected to lead the diggers' rebellion.

Lalor was a civil engineer born in Raheen, Ireland in 1827. He had never previously involved himself in political activity, but 'needs must when Governor Hotham or the devil drives' and conflict like that about to flare on the Victorian Gold fields, is famous for throwing up natural leaders.

It was at this meeting that a Canadian named Captain Charles Ross suggested the Southern Cross as an appropriate design for the flag that was to fly over the struggle and rival the Union Jack as a symbol of the Southland. Two English women worked through that very night to produce the white starred cross on blue background and at 11 o'clock the following morning, the Southern Cross unfurled and snapped in the breeze when it first flew before a crowd of 15,000 diggers on Bakery Hill, Ballarat.

At two in the afternoon that day, Peter Lalor and his two Lieutenants Timothy Hays and Raffaello Carboni, spoke at length to the assembly. They received Governor Hotham's refusal to drop the charges against McIntyre, Fletcher and Westerby from the delegation sent to try and obtain their release, and this prompted a German ex-military man Frederick Vern, to propose a general refusal to pay licence fees.

Timothy Hayes leapt forward and shouted to the assembly "If a man was jailed would a thousand liberate him?" His voice was drowned in the roar of affirmation.

"Two thousand" roared Hayes, and again the air shook as thousands of voices howled their agreement. Hayes again doubled his number to 4,000 and as the miners voiced their solidarity, he bellowed back "Are you willing to die?" Ballarat echoed and re-echoed with the cheers from 15,000 throats.

Volleys of rifle and revolver shots blasted skyward as licences were thrown in heaps and lit and the huge gathering dispersed in good order as smoke from the charred symbols of oppression drifted skyward.

On the following morning, authorities attempted a licence hunt but were met by a throng of diggers ready to resist. Chief Commissioner Rede read the riot act and when none of the miners heeded the warning, troops advanced on the assembled men who scattered before them. Eight were charged with not possessing licences.

At three that afternoon the Southern Cross again flew on Bakery Hill and Peter Lalor knelt with his right hand pointing to the flag and proclaimed "We swear by the Southern Cross to stand truly by each other and fight to defend our rights and liberties". His men replied "Amen."

On Friday 1st December 1,000 men marched to the site of the burnt Eureka Hotel and under the command of the German Frederick Vern, they began enclosing a large area with logs, slabs and anything else they could rake up to construct the Eureka Stockade. A steady stream of diggers came in to join their mates and on that first day of December 1854 Peter Lalor was elected President of the newly proclaimed Republic of Victoria with Raffaello Carboni as Vice President [7].

To preserve security within the stockade, a password was decided on and in commemoration of the only other civil conflict Australia had seen, the words 'Vinegar Hill' were chosen.

Governor Hotham denuded Melbourne of garrisons and pressed his citizens' home militia into service and soon every man who could be ordered or cajoled into a uniform, was marching for Ballarat. Commissioner Rede handed over command of the military to Captain John Wellsley Thomas of the Second Somersetshires who had seen service in India and early on Sunday 3rd December, he marched 100 mounted and 176 foot soldiers into position outside the stockade before sunrise.

Lalor and his men were caught flat-footed by Thomas's strategy as the stockade was lightly manned with many detachments of diggers piloting new recruits in and commandeering arms and supplies. Many others had also gone to attend morning Mass with Father Smyth, a Catholic priest who had firmly allied himself to the diggers' cause and of course there were those who were sleeping off the night before in their shacks and tents outside the stockade.

The diggers' army had made an awful tactical mistake with only 200 men at the defences as Captain Thomas's troops stole into position. Thomas deployed mounted soldiers and police to either flank, and sent sixty-four infantrymen and police in for a frontal attack. The rebels leapt to their defences and began firing, but the advance was rapid and fighting was soon hand to hand.

Casualties began to mount as the American Rangers Revolver Brigade took the main force of the attack. Peter Lalor fell with a bullet to his shoulder and the outmanned and outgunned Eureka forces were put to rout in twenty minutes of savage fighting. One of the last men to go down from his wounds, was an American Negro named John Joseph who by a stroke of luck survived the ensuing prowl by soldiers over the field bayoneting wounded and tearing down the Eureka Flag. The flag's designer, Captain Charles Ross, was fatally injured and stated to several witnesses before he died, that he had been a prisoner for ten to fifteen minutes before he was shot [8].

There were twenty-two digger casualties in the battle. Father Smyth performed last rites on dying men until he was ordered from the field, and in the confusion, some rebels managed to smuggle Peter Lalor to the priest's house where he was hidden to eventually be spirited away to his fiancée's home at Geelong. Thomas's forces received four deaths and twelve wounded. 160 prisoners were held in custody and although most of them were released soon afterwards, a few were held for some months pending treason trials planned by Governor Hotham.

Twelve members of the diggers' forces finally faced court on treason charges. The first was John Joseph the American Negro. Joseph's trial began on 22nd February 1855 and by 27th March he and his eleven fellow accused had all been tried and acquitted [9].

Repercussions of the Eureka Stockade were felt around the world with condemnation of the Crown's action from many countries sending England into damage control. This resulted in Hotham resigning from government at the formation of Victoria's first Parliament where Peter Lalor was elected unopposed to represent Ballarat on the Legislative Assembly of Victoria on 10th November 1855.

The Queensland Shearers' Wars

The Strike of 1891

Australia's greatest conflict involving sheep was not a range war involving cattle and sheep factions like some in America, but a bloodless uprising between 'haves' and 'have nots'

On one side were the monied interests, wealthy squatters and the banks which were backed to an almost shameful degree by the conservative Griffith Government of the day.

The opposing forces were the shearers and shed hands who attempted to gain standard rates of pay and decent conditions by the formation of unions and a show of solidarity.

The old journalists' adage of 'research the facts and print the legend' was certainly adhered to during the civil unrest of the 1890s. Even today the image of the gun-toting red-ragger, harassing peaceful pastoralists springs to mind when people remember the shearers' strikes.

The advice given by Colonel Tom Price, when he addressed his force of Mounted Rifles, was if the order to fire was given he wanted to see no guns aiming in the air and the soldiers' duty was to "aim low and lay them out", has become entrenched in the legend of the Shearers' War.

Many people firmly believe this happened in Queensland during 1891. The 'lay them out' order was in fact issued in Melbourne during 1890 when Colonel Price and his men were called out during the Maritime Strike.

The Shearers' and Labourers' Unions supported the Maritime Strike. Southern pastoralists supported the Shipowners and somehow the catch phrase 'aim low and lay them out' has been attributed to the Shearers' War.

The ultimate ringleader of the pastoralist faction was one George Fairbairn Jnr. who covered his tracks so well, he has been mentioned in passing as no more than a member of the Pastoralists' Special Executive by most historians.

Fairbairn's father, George, was a Scott who migrated to South Australia in 1839. He made a rise as a gold buyer on the Ballarat fields and went on to amass huge pastoral holdings in Victoria and New South Wales during the 1860s [10].

By 1876 he had moved into Queensland and by then was one of the first four men in Australia to own a million sheep. George Fairbairn was semi-retired by 1891 when the strike broke out but his eldest son George Jnr. held the reins of an empire that owned or controlled twenty-eight Queensland stations running over 3,000,000 sheep plus the family's southern holdings.

The most common belief about the cause of the 1891 strike is of pastoralists attempting to cut the shearing rate from £1 per hundred sheep shorn, to 17/6. This in fact happened first in New South Wales in 1886 and was responsible for the formation of some unions, but shearers' unions existed on the Darling Downs and Central Highlands regions of Queensland as early as 1874.

The rate reduction was a repercussion of the plunge in wool prices in 1885 but within a few years, graziers were paying the £1 rate again because shearers began leaving the industry.

The real reason for the 1891 strike was pastoralists and financiers attempting to crush rural unionism in a reaction to the general call out over the Maritime Strikes of 1890.

The Central Queensland Employers' Association was formed at Barcaldine on 15th April 1889. This meeting was proposed at a clandestine pow-wow at Fairbairn's Melbourne mansion earlier that year. George Fairbairn, his trusted lieutenant Francis Murphy, who was the member of the Legislative Assembly for Barcoo, and nine of his managers, made up eleven of the twenty-one present.

Andrew Crombie the 'owner' of *Strathdarr* Station near Mitchell, is credited as being the driving force behind the formation of the C.Q.E.A. Crombie purchased *Strathdarr* for £36,000 in 1882. £35,000 was put up by the Trust and Agency Company of Australia Ltd of which Fairbairn was a Trustee.

One of the first actions of the newly formed pastoralists association was to cut shed hands wages by 5/- a week from 30/- to 25/-. At this time the squatters also entered an agreement with the Carriers Union that allowed a 10% reduction in long distance haulage rates in return for an undertaking by pastoralists to employ only union members. This locked the carriers union into an agreement that effectively hobbled them with regards to supporting the shearers.

On 16th August 1890 the Maritime Strike began in the southern states when the Marine Officers' Association walked off their ships in protest against pay rates. The Queensland Shearers' Union and Queensland Labourers' Union levied their members £1 per head to support the striking seamen. Both shearers and labourers unions were called out on 1st September with the proviso that sheds already shearing be allowed to continue working. With the unions in a shaky financial position due to their support of the Maritime Strike, George Fairbairn decided to make an attempt to break the rural labour organisations once and for all.

By this time the Central Queensland Employers' Association had amalgamated with three other similar groups to form the United Pastoralists' Association and they were more than ready to take the unions on when *Logan Downs* Station held its roll call where available workers presented themselves for possible positions on 5th January 1891. Labour representatives urged strike action and this date is most often quoted as the beginning of the Shearers' War.

The first strike camp was established at St.George with camps at Clermont and Barcaldine being set up soon after. The pro-squatter press of the time had a field day with reports of weapons caches and planned armed rebellion.

Shearers from *Bullamon* station residing in the St.George strike camp were prosecuted for breach of contract when they walked off the job and received a fine of £5 each. An interesting aside to the court proceedings was the fact that prosecuting attorney, Railways Minister Theodore Unmack, who earned the name of 'Ten Bob Ted' when he claimed it was outrageous for workers to be unsatisfied with wages of 10/- per week, was paid £156 for his day's work in court.

Strike-breakers were imported from Melbourne in February. The agreements these men were hired under were later found to be illegal, but they were threatened with imprisonment if they failed to observe their contracts to the letter. Guns and ammunition supposed to be the property of the unions, were seized at Emerald and 400 striking unionists attending a meeting at Barcaldine on 15th February made threats of armed conflict and burnings of grass and woolsheds. Things were hotting up.

The Government's first response was to dispatch an extra sixty police to Rockhampton and order regional police magistrates to draw up lists of potential special constables. Few of the citizens listed had any wish to serve however, and in mid-February Premier Sir Samuel Griffith ordered troops to Clermont.

Thirty men and two officers of the Rockhampton mounted infantry under Major Percy Ralph Ricardo were the first to arrive on 21st February. The first incidence of rioting happened at Clermont but reports on this were all given by pastoralists and their supporters, so the whole thing may well have been orchestrated by the squatter faction.

Seven arrests were made and committed to the Rockhampton Circuit Court. Bail was granted but the men were not released due to a directive that if let out, they were to be immediately re-arrested on other charges. The prisoners elected to stay in custody rather than incur further legal fees for the union.

The trial began on 22nd April under Judge George Rogers Harding who as good as advertised his sympathies when he made the defendants stand at attention in the dock for the entire first day, and half of the following day.

The jury found all seven 'not guilty' but all were re-arrested on various charges and five eventually served a month's imprisonment with hard labour.

Grass fires were started by strikers throughout the sheep country but as 1891 was a wet year, they failed to cause much damage and probably assisted the stations by burning off dead grass.

The first threat of actual armed conflict happened when Major Ricardo and his men escorted strike-breaking, blackleg workers from *Rainworth* Station to Springsure, to catch a train to Ebor Creek near Capella. Other blacklegs from *Arcturus Downs* joined the train at Minerva under an escort of mounted infantry under Sub-Inspector White. The strike-breakers were loaded on wagons at Ebor Creek to continue their journey to *Peak Downs*. A party of between forty and eighty mounted unionists accompanied the cavalcade trying to induce the blacklegs to join the strike.

George Fairbains' brother, Charles, who managed *Peak Downs*, was present when the party crossed onto the stations' freehold. As soon as they were inside the boundary, Fairbairn told Major Ricardo that he was on private property and to order his men to open fire on the strikers. Ricardo refused and when more unionists came up, White ordered his men to fix bayonets. After some verbal sniping by both sides, the union men withdrew ending the Peak Downs 'riot'. Arrests followed later [11].

The Griffith Government's main instrument in the Shearers' War was Colonial Secretary, Horace Tozer. Tozer was blatantly on the side of the graziers even going so far as having the police magistrate at Barcaldine transferred to Muttaburra when he showed some sympathy with the shearers.

The first woolshed burning was at *Maneroo* Station west of Longreach. Shearers in a strike camp at Arrilalah were determined to force blacklegs off *Maneroo* and after two attempts to cross the flooded Thompson River where a striker named John Connell was swept away and drowned, twenty men rode up to the huts at *Maneroo* to find the strike-breakers were being kept at the head station.

On 6th March the *Manaroo* Woolshed burnt down. No arrests were made and many people suspected that the shed which was riddled with white ants and near collapse, had been fired by the owner to collect insurance.

As the strike gathered momentum, more soldiers and police were sent to affected areas. Eight thousand men were in the strike camps and some western towns appeared to be almost in a state of siege. *Milo* station was raided by a mob of men on 15th April, and the woolshed and eighty bales of wool burned. On the same day, the woolshed at nearby *Gumbardo* went up amid somewhat suspicious circumstances. Little was proven but the insurance company refused to pay out either for the shed or the 576 bales of wool that were claimed to be in it.

Five woolsheds were burned in the 1891 shearers' strikes and only the *Milo* shed could definitely be claimed as the work of an organised gang of unionists. *Nive Downs* and *Lorne* sheds may have been fired by strikers working independently of their executive, but they were not destroyed on orders from the unions.

One tradition that has become a distinguishing mark of the Australian Military had its beginnings in the Shearers War of 1891. Troopers of a Gympie Mounted Rifles Unit around Barcaldine began decorating their hats with Emu feathers and the distinctive plumes have survived to this day wherever Aussie soldiers wear the slouch hat [12].

In this almost bloodless uprising there was only one violent death recorded. Two deserting blacklegs, William McLeod and James Ryan, had given themselves up and were given work on *Barcaldine Downs*. On 15th April the same day the *Milo* shed was burnt, McLeod stabbed Ryan to death in his sleep.

Horace Tozer removed two more magistrates accused of sympathy with the unions and used his senior police officers as spies to keep tabs on members of the judiciary throughout the west.

At least thirty men were charged and stood trial for a variety of offences as a result of the Shearers War. Of these thirteen involved in the *Peak Downs* riot received sentences of three years hard labour for conspiracy and rioting.

'Happy Jack' MacNamara received three years for aiding and abetting arson and James 'Shearblade' Martin got two years for sedition. At James Martin's trial in Rockhampton the presiding Judge, George Rogers Harding, threatened to lock the jury up indefinitely until they returned a favourable (to him) verdict.

The other unionists arrested got off relatively lightly with sentences of a few months on a variety of charges.

James Martin would go down in history as a working class hero of the Shearers' War. He had spent most of his working life following the isolated profession of boundary riding and immersed himself in the works of socalist writers like Marx, Bellamy and Nordeau. On 21st March 1891 he addressed a protest meeting at Barcaldine claiming to have a petition in his swag that would result in 10,000 resolute bushmen armed with 10,000 shear-blades making the government take notice of electoral reform.

Martin's words resulted in his arrest and imprisonment on sedition charges, but the union struggles in Western Queensland had not heard the last of 'Shearblade' Martin.

In the wake of the trials a wave of panic washed through pastoralist ranks sending them scuttling to brief lawyers. If unionists could be convicted on conspiracy charges, perhaps the unions could return the favour. Had many squatters not openly bought and stored arms and ammunition? Had not the Pastoral Employers Association raised a private army when R.G.Casey put 140 vigilantes together calling them 'Gentleman Cadets' and outfitting them at a cost of £15 per man.?[13]. In the case of Charles Fairbairn had he not called on unarmed men to be shot?

Luckily for the squatters, sauce for the goose did not prove to be sauce for the gander. The unions' funds were at a very low ebb after maintaining strike camps and further legal action was

James 'Shearblade' Martin, Queensland's most famous revolutionary.

Photo courtesy John Oxley Library Negative no. 67375

out of the question. This lack of funds was the main cause of the breakdown of the 1891 strike.

By May, when the cold weather had set in, life in the poverty-stricken strike camps was well nigh intolerable. Men began drifting away and Horace Tozer sensing he had won, began winding down costly police and military presence in the area.

Heavy rain fell at the end of May and the strikers decided to move to one camp at Barcaldine. Floods prevented the move and once the money ran out, local businesses which had done very well out of supplying the camps, began to withdraw their support.

Despite last-ditch efforts by union leaders to rally the hungry and cold strikers with promises of cash and rations, the strike ground to a halt by mid-June. The strike committee issued its final manifesto on 20th June urging its members to join the electoral roll and continue the fight as members of the fledgling Australian Labour Federation.

The 1894 Strikes

In the aftermath of the strikes of 1891 many of the victorious pastoralists were in favour of immediate rate reductions but cooler heads prevailed. It was rightly reasoned that to rub salt in the shearers' wounds by forcing their pay down further would push even non-union men towards the labour movement.

The 1892 season passed quietly with record numbers of sheep being shorn. This was the year the renowned Jackie Howe shore 321 merino ewes in seven hours and forty minutes with blade or hand shears. Howe set this tally at *Alice Downs* near Blackall on the 10th October.

Howe's record is popularly supposed to have never been beaten, but it has been bettered by a New Zealander. On 13th February 1976, Peter Casserly of Christchurch, New Zealand, blade shore 353 lambs in nine hours at Colin Gallager's *Rangatee* Station near Mount Somers, Canterbury in the South Island[14]. Jackie Howe's record stands alongside this and could surpass it, as Howe was shearing ewes. Although they were reputed to be light-woolled with a large percentage of 'Rosellas' i.e. bare necks and bellies, had Jackie shorn for nine hours, his tally may have been higher than Peter Casserly's world record[15].

Back to the story. At a meeting in July 1893 the Pastoral Employers' Association resolved to cut the shearing price to 17/6 per hundred for hand shearing (blades) and 15/- for machine shearing. Machine shearers were also to provide their own combs, cutters and oil.

Labourers' rates were slashed by 20% and the rations for all were cut to flour, tea, sugar, meat, salt, soda and acid only. This decision was amended when presented to the Annual Meeting of the Pastoralists' Federal Council in November 1893 when it was decided the shearing rate cut would only apply on the Darling Downs while other areas would retain the £1 per hundred rate.

It appears many pastoralists feared more union problems over the proposed reduction. This would seriously affect them in a time when they were enjoying the highest wool prices seen in Australia to date. While there was considerable support for this miserly piece of legislation, only the Darling Downs squatters appeared to be prepared to risk anything that could hold up their annual wool harvest.

Once the new rates were published, the Australian Workers' Union along with the Amalgamated Workers' Union of Queensland and the Australian Labour Federation made the decision on 18th May 1894 to seek an audience with the United Pastoralists' Association. In typical high-handed fashion, the squatters informed the unions they were not interested in talking and on 16th July the UPA decided to once again introduce 'free' or Blackleg labour. The reaction was instantaneous. Strike camps were formed, picket lines were moving through the sheep country and woolsheds began to burn.

The Government published a manifesto called *The Manual for Shearers Disturbances, 1894* to advise local councils how to select and appoint special constables. The handbook was never intended to become public as it was connected with a 'state of things not necessarily unlawful' but was designed to provide guidelines for putting down strikes [16].

The centre of hostilities was further west than the 1891 conflict with most of the action happening around the general area of Winton. A group of men who set up their headquarters a short distance from the main Winton strike camps became known as the 'Amor Push' after their leader John Amor. The push also counted among its ranks the hero of 1891 James 'Shearblade' Martin. Martin had been released from St.Helena prison in January 1893 and had obtained the position of organiser of the Charleville branch of the Australian Workers' Union of Queensland.

The Amor Push made their presence felt at the roll call at *Oondooroo* Station with Amor assaulting a man who came forward to sign on. The following day, the *Ayrshire Downs* woolshed between Winton and Kynuna was set alight and shots were fired at two station hands who attempted to fight the blaze.

The Siege of *Dagworth*

A total of eleven woolsheds were burned during the 1894 controversy, seven of them in the Mitchell area, three around Winton and one in the Warrego. The most publicised arson attack was when the *Dagworth* shed near Kynuna was put under siege by about a dozen armed men, and then set ablaze. The woolshed was heavily fortified by log barricades and defended by special constables and station employees.

The attack took place shortly after midnight on 2nd September and although many shots were fired, no one was hit. Considering the fact both attackers and defenders were bushmen, well versed in the use of firearms, it would appear nobody really wanted to shed blood and the gunplay was intended to make the point that both sides were serious.

Police and station men at the fortified Dagworth woolshed in 1894 shortly before the shed was burned by striking unionists. Fourth from right in white with belt pouches, is Bob Macpherson, The Squatter from 'Waltzing Matilda'.

Photo courtesy John Oxley Library Negative no.50580

One of the strikers crept in with a kerosene bottle and fired the shed then the attackers withdrew. The battle of *Dagworth* Station was one of only three times in Australia's history, when organised opposing white forces engaged in civil warfare with firearms [17].

The following morning Bob Macpherson, the *Dagworth* manager, led a police party to the strikers' camp at a nearby waterhole where they found the body of a man who appeared to have committed suicide by shooting himself. Unknown to the strikers until the fire started, 140 ram hoggets were in the shed and they perished in the blaze. Nobody wanted this to happen and the dead man who was known as Samuel 'Frenchy' Hoffmeister is believed to be the man who set the fire then killed himself in a fit of remorse over the death of the sheep.

This incident was the inspiration for the song '*Waltzing Matilda*' by Banjo Paterson who was in the area shortly afterwards not on a casual visit as most people believe, but in his capacity as a lawyer. Paterson was sent to broker a truce between the shearers and squatters. He was perfect for this job as apart from his qualifications as a solicitor, he commanded great respect among bush people for his skill as a balladeer and also for his masterful horsemanship.

The strike was declared 'off' by the Longreach branch of the Australian Workers' Union of Queensland on the 10th September and the other two branches followed by the end of the month.

Before the strike ended many pastoralists had agreed to shear under union rules and the potential ferocity of the 1894 struggle showed the squatters and their backers that the unions were prepared to back their beliefs to the hilt.

The Winton and Kynuna Unionists were the bitter end holdouts in the 1894 dispute and the role Andrew Barton Paterson played in reaching agreement has gone largely unnoticed. It was Banjo Paterson who bought the two factions together in the Kynuna Hotel now known as The Blue Heeler, where they settled their differences over a case of lukewarm champagne.

With feelings running at as volatile a pitch as they were between shearers and pastoralists in 1894 it was a marvellous effort on Paterson's part that he persuaded them to have a drink and talk out their differences. Australia will always revere The Banjo for his marvellous poetry but perhaps he should also be honoured as a negotiator and man of peace.

The Aftermath

It was not until 1896 that official action was taken over the burning of the *Ayrshire Downs* Woolshed. David Bowes who had been charged with unlawful assembly after the 1891 *Peak Downs* riot, James 'Shearblade' Martin and two other men named Edward Cowling and James Loyola were arrested and charged with arson. Bowes, Loyola and Cowling received ten years hard labour each for arson and 'Shearblade' Martin was sent down for fifteen years on the same charge ending the career of Queensland's most notorious revolutionary.

The shrine of the Queensland shearers' strikes is the roll-up tree site in Barcaldine which is given as the birthplace of the Australian Labour Party. Shearers in Australia's eastern states became some of the staunchest union men on earth following the strikes. It is only in relatively recent times when the wide comb dispute of 1983 plunged the shearing industry into chaos once again, that unionism faded from the scene,

An old almost forgotten piece of doggerel composed by some anonymous union supporter sums up the attitude to non-union labour in the sheep country of Australia's eastern states from the 1890s until the 1980s.

You Bastard
Why call me a bastard? What have I done?
Scabbed, you bastard in ninety one.
And what is more, you son of a whore,
You scabbed again in ninety four.

Other Australian Conflicts

Australia experienced a few other battles with non-indigenous participants in the early years of white settlement probably the most publicised being the Lambing Flats Riots.

Conflict between Chinese and Europeans on Australia's gold fields was a common thing and flare-ups occurred on a few occasions. White diggers saw the celestials as a threat due to their large numbers and at Lambing Flat Gold Field now known as Young in New South Wales, racial feelings boiled over into violence in December 1860.

Anti-Asian feeling had been fermenting since the arrival of large numbers of Chinese early in the rush and following a roll up call, bands of white miners attempted to drive them off the gold field. Troops were called in from Sydney to restore order and in the subsequent clash between troops and diggers, one death and several woundings resulted. The riots were of short duration and a ring-leader, William Spicer, was jailed for his part in the affair [18].

Another conflict involving white and Chinese miners occurred at a new gold field near the head of the Margaret River, Western Australia in 1880.

A new field had been opened up there by two men named Tennant and Kirk. The enterprising diggers piloted 1,000 Chinese miners to the area at a cost of £1 per head. Tennant and Kirk were well pleased with the £1,000 they pocketed and the field soon became known as a rich one.

Miners rushed in as they did to all new gold finds, and a party of prospectors followed a rich lead right into the Chinese camp. The entire area was soon pegged but the Asians refused to move. Diggers were sinking shafts all over Chinatown and when a miner named Fred Stone struck a rich seam, the Chinese miners present, saw the wealth in the paydirt. In a direct reversal to the happenings at Lambing Flat, the Asians called a roll-up of their own.

Fred Stone was badly hurt in the resulting melee as Chinese tore into whites with sticks, stones, shovels and waddies. Policeman August Lucanus, put a hurried posse together and he and his men charged the rioting miners on horseback scattering them over the gold field and striking them down with gun butts and the flat of swords [19].

As usual in Australian race relations during the 19th Century, legal opinion was on the side of Europeans, and the incident became just one more gold fields squabble that whites won.

The Palmer River Gold Field in North Queensland played host to a Tong War when Cantonese miners suspected a contingent of Pekinese had been raiding their tail race at night. The Cantonese picked their best hatchet men and after lightly doping them with opium, which was supposed to make them more warlike, sent them creeping down on the night raiders who were furtively filling baskets with pay dirt.

The Cantonese charged like banshees and when the clubs, tomahawks and shovels, that comprised their armament, stopped rising and falling, all but one of the Pekinese gold thieves was dead [20]. At dawn the next day, the Pekinese miners led a retaliatory raid and in a half hour of vicious fighting, both sides sustained heavy casualties.

Police under Warden Sellheim came from Maytown but over the next few days fighting broke out repeatedly soon after the troopers broke it up. Eventually about thirty ringleaders were jailed and a rather delicate peace was achieved.

A third faction entered the game when a group of Macao men took advantage of hostilities to move in on a patch of ground, and determined to hold it, built a stockade and stood guard over their newly claimed territory.

Details of the events are sketchy at best, but it would appear the other Asians joined forces to attack the Macao men's fort and around 2,000 Chinese put the place under siege. The Asians were apparently better hand-to-hand fighters than marksmen and although a lot of firing went on, there were only a few casualties in the actual fighting. Many combatants were wounded however, and an eyewitness account of the event claims at least 200 died of wounds and were secretly buried[21].

For a time white miners who were vastly outnumbered by Chinese on the Palmer's alluvial field, thought the Asians meant to drive them from the field, but once they realised they were not targeted, some took up vantage points on nearby high ground to enjoy the show and others took advantage of the diversion to rob Chinese camps.

After a further three days of hostilities, Warden Sellheim's police surrounded the combatants and while the riot act was being read, white miners collected all the guns and destroyed them. The fight became known as the Battle of Lukinville named after George Lukin, Under Secretary for Mines. It was probably the bloodiest skirmish on the Palmer Gold Field, which was also notorious for clashes between miners and cannibal tribes of the north with sinister reprisal raids that invariably followed Aboriginal attacks.

Civil Conflict in America

The Mountain Meadows Massacre

One of the most brutal conflicts between Europeans to happen in North America sprung from religious roots and became the worst massacre of whites in Western history.

In the mid 1840s Brigham Young and his fellow 'saints' as followers of the book of Mormon were known, made their way into the wilderness to escape the persecution their religion received in the settled regions of the United States. The Mormons carved out a kingdom for themselves around Utah's great Salt Lake Basin and became extremely sensitive about non-Mormons or gentiles as they called them, crossing their boundaries.

To be fair to the Latter Day Saints, they had been forcibly evicted from north western Missouri where fourteen of their number died in the Haun's Hill massacre of 1838. They were again thrown out of Nauvoo, Illinois where their founder Joseph Smith, was murdered in 1844, and those who endured the privations of the

Great Trek, as they followed new leader Brigham Young to Utah, would naturally have fiercely resented any outside interference.

Near the end of August 1857, an emigrant wagon train of non-Mormon settlers from Arkansas and Missouri, passed through Salt Lake City on its way west [22]. The Mormon faith was in a state of uproar over a Federal Army expedition into Utah that year following the announcement of the church's belief in polygamy. War with the U.S.Army was averted but the hysterical reaction to their borders being breached by the wagon train had tragic results.

There is no doubt the Mormons were determined to defend their promised land to the last breath. They had their own police force led by two unsavoury characters named William Hickman and Porter Rockwell. Despite the fact that these Mormon enforcers did not practise turning the other cheek, it is still very difficult to reconcile the happenings of 10th September 1857 at a spot named Mountain Meadows, with any Christian teachings or beliefs.

The wagon train was allowed to depart Salt Lake City unmolested and rolled towards California through south-west Utah. On orders from Church leaders, a large party raised and led by Bishop Higbee, President J.C.Haight and Brigham Young's adopted son Bishop John D. Lee, disguised themselves as Indians, and bedecked with paint and feathers, swept down on the immigrants [23]. If the Mormons were hoping for a quick victory they were disappointed as the settlers barricaded their wagons and successfully held the bogus redskins at bay for five days. A new strategy was needed.

The attackers retired from sight on the fifth day and after removing their disguises, they rode down again under a flag of truce representing themselves as rescuers who would negotiate with the Indians with whom they claimed to be on good terms, for the safe passage of the emigrants.

After a period of discussion during which appearances were kept up by messengers riding to the top end of Mountain Meadows to 'parlay' with the non-existent savages, the settlers finally agreed to a supposed ultimatum from the Indians to march back the way they had come on foot leaving all their goods behind. The immigrants were down to their last few litres of water after the five day siege and the idea of walking back to Salt Lake City with an armed guard seemed preferable to certain death by Indians or thirst.

The Missourians and Arkansawyens laid down their arms and marched away with their escort of Mormons. They were about two kilometres from the abandoned wagons when the 'Saints' opened up on them killing nearly all the men in their first volley and pouring fire into the helpless party until 118 men, women and children lay scattered in the grass of Mountain Meadows.

In the confusion, two men escaped and for several desperate days they fled 250 kilometres south into the desert before they too were overtaken and slaughtered. Of the entire party only seventeen children under the age of seven were spared, and it is due to testimony by some of them, plus other deserting Mormons sickened by the affair, and from Paiute Indians who were supposed to take the blame, that the shameful story of the Mountain Meadows Massacre was heard.

A Utah Federal Judge named Cradlebaugh attempted to prosecute the perpetrators but it proved a lost cause and nobody was brought to justice for the atrocity. The events at Mountain Meadows refused to die, however, and almost twenty years later, Bishop John D.Lee was tried for his part in the massacre. A jury of five gentiles and seven Mormons became deadlocked and a mistrial occurred.

The public outcry that followed when it appeared Lee would again escape retribution, convinced Mormon Church leaders they would have to offer someone to justice for the crime. Lee was retried in September 1876, found guilty, excommunicated and sentenced to death. A bizarre ending to John D.Lee's life occurred on 23rd March 1877 when he was executed by a five-man firing squad while sitting on his own coffin at the Mountain Meadows massacre site [24].

The Johnson County War

The conflict that raged over Johnson County, Wyoming in 1892 was a last ditch attempt by big cattlemen to drive out small holders and maintain an era when large scale graziers lived like virtual kings in their prairie principalities.

Times were a-changing. The crash of the beef market in 1885, the movement of homesteaders into country that was only safe for large monied concerns due to the threat of Indians in earlier days of settlement, and the spread of wire fencing, all contributed to the end of the old open range days. Hundreds of cowboys were thrown out of work in the mid-1880s.

Many took up homesteads and tried to make a rise as small cattlemen. Cattle were virtually there for the taking, roaming unbranded on ranches their owners had walked away from. To be sure, prices were low but the stock raisers' faith in the belief that 'people have to eat', prompted former employees of the big outfits to try and take up land of their own and build for the future.

Not all of the free-spirited range riders wanted to settle down to the monotony and hard work that goes with building a property from scratch. In Northern Wyoming adjacent to Johnson County was a remote valley with only one trail in or out where a narrow gorge cut through 300 metre high red cliffs named Hole in the Wall. A handful of men could defend this natural fortress against an army and 'The Hole' soon became a refuge for the long rope brigade who used it as the northern staging station for stolen horses and cattle that could be moved down 'The Outlaw Trail' clear to Mexico.

A loose alliance sprung up amongst the inhabitants of Hole in the Wall with many groups and individuals pursuing their various brands of longhorned larceny. The best known was a group called 'The Red Sash Gang', who distinguished themselves by wearing a scarlet cummerbund around their waist.

The large ranchers of Johnson County lumped the rustlers together with smallholders or 'Nesters' as they contemptuously labelled them, and made plans to use the cow thieves as an excuse to declare war on anyone who encroached on grazing they regarded as theirs by right. To 'Nesters' and Rustlers alike, the big concerns were the common enemy. By law, an unbranded beast not running with its mother was the property of anyone who could catch it, but the big operators had long discouraged any mavericking unless they did it themselves.

The big ranchers had plenty of political clout with state and federal government but on their home ground, they were outnumbered, and soon members of the opposing forces held all the local offices.

The ranchers first move was to persuade the state government to enact a Maverick Law that made all stray cattle the property of the Wyoming Stock Growers' Association to be distributed amongst them as they saw fit. Their second move, was to hire a team of stock detectives to infiltrate the opposing factions and apprehend cattle and horse thieves.

It enraged the smallholders no end to find that the sale of cattle appropriated under the new law was financing the detectives and after some of their numbers were lynched including a woman named Ella Watson, there was no one in Johnson County who had not taken sides with one faction or the other.

Ella Watson was a homesteader and a prostitute. A practical girl, she stocked her land with cattle she had taken in trade for her favours and while it was probably not regarded as a very moral way to enter the cattle business, if any of Ella's stock were stolen, the thieving was done by her paramours and not her.

Ella and a fellow homesteader named Jim Averill who had established a store on his land and owned not one head of cattle, were placed in a wagon one hot day in July 1889. The wagon was parked under a tree while nooses were fitted then the unfortunate victims were pushed off the wagon bed and both died from strangulation.

Members of the lynching party were moved out of the country, one as far as England. Of the three people who witnessed the crime, one a crippled boy, was taken away by the lynch mob and conveniently died a few weeks later. The other two named Cole and DeCory were never seen again. A persona was invented for Ella Watson and the lowly prostitute posthumously became 'Cattle Kate', scourge of Johnson County and Bandit Queen.

It would be tedious to recount every incident that occurred over the following two years, but by the end of 1891, Johnson County was ready to erupt.

Early in 1892 a hard visaged group of cattle barons met in a private room at the exclusive Cheyenne Club. The battle lines were drawn for an invasion of Johnson County to be immediately followed by a wholesale assault on Hole in the Wall. The Stock Growers' Association drew up a death list of potential victims and Major Frank E. Wolcott, an ex-army officer and rancher, was selected as the man to command the invasion [25].

A squad of twenty-four Texas gunslingers was imported to swell the numbers of the participating cattlemen and their trusted minions and on a snowy morning on 7th April, fifty-four men and horses complete with supply wagons that would follow the advance party with food, tents, bedding, extra arms and ammunition, disembarked from a specially hired train at Casper, Wyoming. The invasion of Johnston County was under way.

The first strike planned was Buffalo, the county seat, and its Sheriff, Red Angus, was one of the first names on the death list. The raiders had cut all telegraph lines to prevent discovery and anybody met on the way was forced to accompany them so the news of the mission did not leak out.

At the Tisdale Ranch near the corner of Natrona and Johnson Counties, the invaders rested, ate and made ready for the push to Buffalo, but while they were there, a scout called Mike Shonsy brought the news that rustlers were at the nearby K.C.Ranch on Powder River. The owner of the K.C. was a cowboy called Nathan 'Nate' Champion whom history has recorded from several points of view.

To the Wyoming Stock Growers' Association, he was a cattle thief and this is supported by the Pinkerton Detective Agency whose files describe him as 'The King of the Rustlers' and the leader of the Red Sash Gang [26].

To members of the settler faction, he was known as a hard working cowboy turned small rancher who came from a good Texas family and was active in district affairs. If 'Nate' Champion was the outlaw chief he was made out to be, it is highly unlikely he would have openly made his home at the K.C. instead of being safely tucked away at Hole in the Wall.

The raiders had a score to settle with Champion. One of the Association's detectives, Frank Canton, who was part of the invading force, had attempted to arrest him the previous November. Before dawn, Canton and three other men had kicked in the door of a shack where 'Nate' and his friend Ross Gilbertson were asleep, firing shots at the figures in bed. Champion pulled a gun from under his pillow and began blasting causing Canton and his accomplices to practically get jammed in the door in their haste to retreat. The would-be assassins vacated Champion and Gilbertson's dwelling on foot with two of them wounded, abandoning their horses and gear. The story had been well tossed around bringing considerable embarrassment to Canton and his henchmen Tom Smith, Joe Elliott and a hired gun named Coates.

The raiding party made the decision they would detour to the K.C. on their way to Buffalo and get Champion and his partner Nick Ray who were both on the hit list they carried. On the morning of 9th April the raiders stole into position around the cabin that served as the K.C. Ranch headquarters in darkness and waited for shooting light.

At what Australians would call Piccaninny Daylight (pre-dawn), an old trapper named Bill Walker who had been staying at Champion's ranch, left the cabin with a bucket heading for a water hole seventy metres away. Walker was quietly taken prisoner and when he did not return, his trapping partner Ben Jones was also taken when he came searching for his friend.

There was no effort made to capture Nick Ray who was the next inhabitant of the cabin to step outside. He fell, shot in the head and body from the ragged volley that cracked out and 'Nate' Champion leapt through the door revolver blasting to wound a member of the attackers and drag Ray back inside.

The rustlers who were going to be casually eliminated en route to Buffalo proved a lot tougher proposition than the raiding force had reckoned on and a siege developed that was still underway when a settler named Black Jack Flagg and his stepson passed at three in the afternoon. Flagg was on horseback and the seventeen year old youth driving a wagon.

The raiders opened fire at a distance with the intention of preventing them spreading the news but the shots went wild and after slipping a team horse from harness for the youth to ride, man and boy galloped away unharmed. The word was out. Champion had to be got and got quickly. The raiders rolled Flagg's abandoned wagon piled high with sticks and hay, against the cabin using it to shield themselves from shots, and set it alight.

Tensely the raiders waited rifles cocked, as flames engulfed the cabin. When it seemed that no one could be alive in the inferno, 'Nate' Champion suddenly burst out of the smoke holding a Winchester and blazing away with his Colt, to fall victim to the hail of bullets that awaited him.

It was later discovered 'Nate' had found time to scrawl down a record of Nick Ray's death from his wounds and his own final desperate hours. The document still exists as a treasured piece of Western memorabilia, but the name of an adversary Champion recognised, probably Frank Canton, was carefully scratched out when the paper was taken from his body.

The invaders made all haste and by way of ranches where they could change horses, they made a night ride for Buffalo. They reached the T.A. Ranch on Crazy Woman Creek just twenty-two kilometres from the town, to be met by Phil Dufran, an Association man who had been planted in the area as a spy.

When Black Jack Flagg and his stepson brought the news of the attack on the K.C.Ranch, Sheriff Red Angus had been far from idle. Reports say he rode almost two hundred kilometres in fourteen hours and raised a force of over one hundred men who as Phil Dufran reported, were even then riding hard towards them.

Major Wolcott called a hasty meeting. It was now realised protection was needed from the settlers and rustlers who had so unreasonably decided to fight back. The hunters had come to the realization they were now the hunted, and they began a frenzied burst of activity to dig trenches and build barricades out of timber stacked there ready to construct a house.

The impromptu fort was hardly thrown together when the enemy appeared and invaders began to get a taste of the treatment they had dished up to the inhabitants of the K.C. The Johnson County settlers' forces were directed by Sheriff Red Angus, the Rev. M.A.Rader, a Methodist preacher, and an old time scout and Indian fighter called Arapaho Brown.

In a carbon copy of the other side's ploy at the K.C., a moveable barricade and torch was made from a wagon and as it began its journey towards them, the now thoroughly chastened invading force realised they could be facing the same fate as Ray and Champion. However, in true western tradition when things look blackest, the plaintive note of a bugle echoed over the battlefield and three troops of cavalry trotted up to rescue the invaders who had planned to 'clean up' Johnson County.

The timely arrival of the U.S.Army saved the Association forces in the nick of time and ranchers and hired gunslingers alike were more than happy to be placed under arrest. By eventual consent between all parties, no one was tried for the crimes although the raiding party was held at Cheyenne while the settler faction vainly tried to have the case tried in Buffalo. Cattlemen's friends in high places made this a forlorn hope and eventually the raiders were released.

Out of the fiasco of the Johnson County War came at least a better understanding of the needs and rights of all classes and never again would large property owners literally wield the power of life and death.

The Tonto Basin War

Probably the most ferocious range war ever fought in America began in Arizona's Tonto Basin in 1887. The country in the lea of the Mogollon Range had been occupied since 1880 when pioneer rancher John Stinson drove a herd onto the grazing of a wide watershed known as Pleasant Valley. By 1887 there were more cattle operations established, the biggest of which were Stinson's, 'The Hashknife' run by the Aztec Land and Cattle Company, and the Graham Ranch.

The Tonto Basin cattlemen established a deadline for sheep and served notice that no woollies would be tolerated on their side of the Mogollon Range. Deadlines against sheep were fairly common in the old West one of the most famous being set by Charlie Goodnight in 1876 when he settled the Palo Duro Canyon and reached an agreement with sheepmen to keep his herds off the Canadian River country if they would accord him the same consideration in the Palo Duro.

Pleasant Valley's pioneer John Stinson began to feel a little hemmed in with no more vacant land around him and he reached a somewhat shady agreement with three of the Graham family – Tom, John and Billy – who were reckless young cattlemen with a taste for action and excitement.

It seems the Grahams were working on Stinson's behalf to intimidate smaller ranchers encouraging them to leave Pleasant Valley and allow Stinson to expand his holdings. The problem for the Grahams was that one of their targets, the Tewksbury Clan, was extremely hard to scare.

John D. Tewksbury was the family patriarch. He was a Boston Yankee who had been through the turbulent years of the California Gold Rush, married an Indian girl who died after giving him three sons, and eventually moved to Arizona. Shortly after arriving in the Tonto Basin country John Tewksbury married an Englishwoman named Lydia Crigler Shultes giving his three sons John Jnr, Jim and Ed, a stepmother. Shortly after the family settled down to work their property, old John Tewksbury died [27].

The Tewksbury boys were lean, Indian dark, backwoodsmen and like their father, they had few peers at tracking, scouting and the use of firearms. No one has been able to accurately establish just what triggered the Graham Tewksbury feud, but it is known there was bad blood between them before Tom, John and Billy Graham began crowding the half Indian family on John Stinson's behalf.

Accusations of rustling had been made by both sides, and when the Tewksburys rejected an offer for their land from Stinson delivered by the Grahams, it deepened animosities. However, if relations were strained then, it was nothing compared to what followed.

In early 1887 a cowboy made a hasty trip in from the valley's northern edge with a report that made cattlemen begin checking their weapons. The Tewksburys had driven sheep over the Mogollon Rim and onto rangeland claimed by the Grahams. The sheep were the property of two brothers P.P. and W.A. Daggs, a rather appropriate surname for sheepmen, but if the Tonto Basin ranchers saw the irony of it they didn't appreciate the joke.

The Daggs had offered the Tewksburys a share in the flock's increase if they would take them across the Mogollons to pasture and Ed, John and Jim Tewksbury were hard nosed enough to fly in the face of the cattlemen's anger. A formal warning was sent to the sheepmen and when that was ignored, a Navajo Indian herder, who worked for the Daggs, was shot in February.

Soon after that, a group of cowboys ran three shepherds to ground on a rocky hill and held them there at gunpoint to watch their flock being slaughtered. The remnants of Dagg's sheep were pushed back out of the valley by the end of the month leaving the Tewksburys to plot a revenge that was only a few months coming.

The unexplained disappearance of Mark Blevans – head of a family of cattlemen – in July was laid at the door of the Tewksburys and by then almost every person in Pleasant Valley had taken sides.

On 10th August a party of eight men led by Hampton, one of the sons of the missing Mark Blevans, went looking for the Tewksburys. They were found at a place known as the Old Middleton Ranch on Cherry Creek and when the powder smoke cleared and the echoes of gunshots died away, Hampton Blevans and a man called John Paine were dead, four others were wounded, one seriously, and the Tewksburys, who had all escaped without a scratch, watched the survivors ride for their lives.

Despite the victory, the Tewksburys knew the numbers they faced and that retaliation was inevitable. They forted-up on a high rocky bluff with a commanding field of fire, and awaited developments.

For a full day the opposing factions conducted a long-range rifle duel and after night fell and the moon came up, Jim Tewksbury carefully picked his way down to a spring to refill their water canteens. One of the attackers had slipped up on the water carrier, and at a warning shout from another Tewksbury, Jim pulled the trigger of the rifle he was holding over his shoulder with the muzzle pointing back.

The Tewksburys were renowned rifle shots but not even the best sharpshooters can pull that trick off without using a mirror. We must assume it was no more than a fluke when the bullet found its mark in the thigh of the luckless stalker who bled to death while the Tewksburys searching rifles prevented any rescue attempts.

The Tonto Basin War was common knowledge by this time and Sheriff William Mulvenon of Yavapai County obtained murder warrants for the Tewksburys, but when he scoured the country with a large posse, the wily frontiersmen were nowhere to be found.

Within a few days the Graham family received their first casualty. Billy Graham was on his way home from a dance in Phoenix on 17th August when he was shot from ambush and fatally wounded. Many years afterwards, it was found his assassin was a sheepman and deputy Sheriff named James D,.Houck who was a Tewksbury supporter. It has never been established whether Houck acted on his own initiative or in collaboration with the Tewksburys, but there was no doubt in people's minds at the time. The Grahams were mad with rage and they enlisted the help of Mark Blevans' sons who were convinced their father had died at the hands of the sheepmen and knew their brother Hampton had.

The three Blevans brothers, Andy, Charles and Sam Houston who were all holders of bad reputations particularly Andy, became the Grahams' deadliest allies and after the Tewksburys had been hunted up and down Pleasant Valley, they were finally run to earth at the Tewksbury Ranch.

Seven men and three women were at the ranch when the cattlemen laid their ambush early on the morning of 2nd September. John Tewksbury and a man named Ed Jacobs had gone out early to bring up some horses and on returning they walked unknowingly into the cowboy rifles. The pair was shot where they stood by Andy Blevans and another day-long siege began with the sheepmen forted in their cabin and the Grahams, Blevans and their followers sniping from the rocks and brush.

During the afternoon a mob of the Tewksburys' pigs appeared and began feeding off the bodies of John Tewksbury and Ed Jacobs. The cattlemen laughed and carried on with the fight. As far as they were concerned, a damn sheepman deserved to be fed to hogs, but old John Tewksbury's widow Lydia proved she had inherited a good dose of the British bulldog from her English forebears.

With a shovel over her shoulder, the game woman walked straight out of the cabin, drove off the swine and buried her stepson and Ed Jacobs in shallow graves.

Not a shot was fired or a sound made as the brave lady buried her dead and after she returned to the dwelling, the attackers pulled out and left the ranch.

The Tewksbury faction got an unexpected bonus a few days later when all but one of the Blevans family died at the hands of Sheriff Commodore Perry Owens in nearby Holbrook. The story of Owen's deadly duel is told in chapter 2 volume 1, so we will not go into detail here. The loss of their most warlike allies did not hold the Grahams back and three days later, they again attacked their enemies who were at a campsite on Cherry Creek. The cattlemen galloped into the sheepmen's camp early in the morning but the Tewksburys simply sat up in their bedrolls and started shooting.

The attackers' marksmanship was noticeably affected by trying to fire from running horses and in a few seconds cowboy Harry Middleton was dead, another, Joe Ellenwood, was wounded, and the Grahams were in retreat.

On 11th September, Sheriff Mulvenon once again led a large force into Pleasant Valley hoping to end the War but they were met by the crafty deputy and Tewksbury partisan James Houck, who convinced Mulvenon the sheepmen were innocent victims, and the entire feud was orchestrated by the Grahams.

The Sheriff and his posse laid an ambush and on 21st September they shot down Charles the last surviving Blevans, and John Graham. Arrests were made from both factions but as giving testimony against either side was an uncertain proposition, no witnesses bothered appearing.

From early 1887 when the bloody conflict erupted, the Graham Tewksbury feud dragged on until 2nd August, 1892 when Ed, the last surviving Tewksbury shot Tom, last of the Grahams. Jim Tewksbury had died of illness while the vendetta was still going on and Ed was convicted of Tom Graham's murder. He managed to obtain a new trial on appeal and the case was finally dropped ending one of the bitterest range wars in the history of the West.

North America saw many other disputes between non-indigenous Americans flare into open warfare - Canada's Metis uprising; the Lincoln County War in New Mexico that drew national prominence to William Bonney known as Billy the Kid; and the Sheep War of Wyoming's Ten Sleep country, to name but a few. Probably the last notable events of earlier times occurred when gang warfare raged during the prohibition era of the 1920s and gangsters like Al Capone, John Dillinger and Pretty Boy Floyd were active. Over 500 gangland murders took place in the torrid decade of the 1920s. The Saint Valentine's Day massacre of 1929, when seven of the O'Banion gang were gunned down at the SMS Cartage Company garage on North Clark Street, Chicago, by opposing interests, was the most notable clash.

Sheep Versus Cows

Relations between sheepmen and cattlemen were always a lot more cordial in Australia than America. In Australia most of the pastoral areas were pioneered by sheepmen and gradually the land itself determined which species would graze it. Cattle that could walk to market took precedence over sheep in the more remote regions where the problems of overlanding an inanimate cargo like wool, presented huge problems. Other areas proved climatically unsuited to sheep and many grew grasses with seed that ruined the fleeces.

Gradually a balance was achieved and throughout Australia's history, sheep, cattle and their respective herders, have existed in relative harmony. By the very nature of their job, Australian ringers, like American cowboys, have always been a bit contemptuous of the slow plodding sheepman. There is plenty of slow plodding attached to cattle work as well, but a bit of dash, speed and danger is always present when handling big mobs of semi-wild station cattle.

Many stockmen worked with both sheep and cattle and bush workers were often as adept with shearing gear as they were with whips and branding irons.

The cattleman's scorn of the sheepman was entrenched in both continents, however, and to this day in Australia's or America's cattle industry it is fighting talk to call anyone a 'sheepherder'.

A good example of the old time cattleman's attitude to sheep and the fences used to contain them can be found in the final two stanzas of Breaker Morant's poem 'Nebine Dick'.

> *Stock-keeping days shall last your time, until you have commenced*
> *The long trip-over Jordan-where the country isn't fenced-*
> *You'll make a bee-line for above, and Nebine Dick, sure then*
> *You shall be welcomed into camp-by angel cattlemen!*
> *And those good Panic colts you rode, down here in 'cattle days'*
> *Shall run on Plains of Paradise, where no Merinos graze,*
> *You'll ride once more, rejoice to find as find- you will, no doubt-*
> *St. Peter has one fence up there- to shut the sheep-men out!*

When sheep began to invade American ranges that had been claimed and held by cowmen, there was no question of amiable relations at first. Sheep and their herders were killed. Waterholes were poisoned, and some brutal range wars were fought over which species would graze the land.

It has since been proven that both animals can exist in an harmonious balance and actually aid pasture management by cross grazing the same area, but in the old West, sheep were believed to ruin land and foul water sources.

Sheep are definitely more destructive grazers than cattle and in the days of open unfenced range, there was a certain amount of truth in the cowman's claims. Fencing and management techniques have enabled graziers to better care for the land in modern times, but both continents contain ruined dust bowl regions that are a legacy of overgrazing by sheep.

WOMEN OF THE WILD WEST

The Paroo may be quickly crossed – the Eulo commons bare
And anyway it isn't wise, old man to dally there.
Alack-a-day far wiser men than you or I succumb
To women's wiles and potency of Queensland wayside rum.

Breaker Morant

Frontier Females

From indigenous partners who served the needs of lonely male pioneers and salt of the earth wives and daughters of battling settlers. From matriarchs of pastoral dynasties and female outlaws, to frontier femmes fatales, women played a huge role in the frontier history of Australia and America.

Some worked themselves quietly into early graves playing the traditional role of wife and mother as they supported men who dreamed of a better life in a new country. Others took the frontier on, on their own terms, and built legends that have lived down through the years.

One of the great entertainers and Australians of our time, Ted Egan, celebrates the contribution made to the early pastoral industry by hundreds of nameless Aboriginal women in his ballad *The Drover's Boy*, about a girl taken from her tribe and forced into concubinage with a white man.

In early Australia an undesirable variety of slaver existed known as Lochinvars after Sir Walter Scott's poem of the same name about a Knight who carries off a fair maiden at the altar under the nose of her new husband.

Scott's poem creates a dashing romantic image for its central character, but the Lochinvars of the Australian bush were no heroes. Their unpleasant trade often involved shooting Aboriginal men who resisted the forcible abduction of their wives and daughters captured and sold to drovers or taken out of their tribal lands and purchased by owners of newly taken runs.

In early times indigenous females were regarded as better and less troublesome stock boys than their male counterparts and there was of course the added bonus of their attraction as bed partners. Some of these unfortunate women developed great loyalty and affection for their white masters. Their bush skills of tracking as well as finding water and grazing, were of inestimable value.

Unfortunately many were passed from hand to hand with no more consideration than the horses they rode and were sometimes abandoned alone and far from their homelands when a white bride became available.

Similar happenings occurred in America particularly in the days of the Mountain Men and the American Squaw Man had his parallel in Australia, the Combo.

There is not a lot of entertaining copy in the virtuous wife and mother who quietly devoted her life to running a home and family and it is usually the women who stepped outside the bounds of traditional values who go down in history.

Some ladies of the early pastoral scene were remembered for their less than charitable attitudes. Mrs. Buntine, wife of early Gippsland, Victorian squatter, Hugh Buntine, is reported to have hated Aborigines with such a passion she amused herself by chasing them one at a time into the sea on Ninety Mile beach on horseback helped along by liberal applications of the Lady's stockwhip. Mrs. Buntine would then patrol the beach forcing the poor wretches back into the sea by slashing them mercilessly and forcing her victims with her horse if they tried to come ashore until exhaustion took over, and they drowned [1].

This story at its source has a whole group being herded into the sea, but it would take a quick horse or a slow tribe and the entire concept is blatantly impossible. But one on one, a bully on a horse could execute another person with ease.

Another matriarch of a well known pastoral family, who was widowed in middle age, ran several properties in the late 19th and early 20th Century and established her own breeding programme for stock riders. The lady in question would arrange matings with Aboriginal women and her Chinese cook. She would then breed her own sons to the resulting female half-castes. The woman was very proud of her 'line' and was prone to bragging that she bred the smartest niggers in the north [2]. Truly a delightful human being!

Frontier boom towns in both lands had their share of Soiled Doves. Longreach in Queensland in the heyday of the wool trade, saw its predominantly male population's needs catered to by a selection of fallen angels who went by such interesting titles as The Scrub Turkey, The Two Bob Touch, The Faded Flower and The Long Handled Shovel [3].

Kansas trail towns where Texas cattle walked to the railheads were home to ladies of the night with sobriquets like Big Nose Kate, Hambone Jane, The Galloping Cow, Velvet Ass Rose, Little Gold Dollar and Squirrel Tooth Annie [4].

One of America's wilder cow towns, Ellsworth, even had its own version of Lady Godiva. A saloon girl known as Prairie Rose made a $50 bet with a cowboy she would walk naked down the main street. The real Lady Godiva, wife of Leofric Earl of Mercia 1040 – 1080 AD, performed her unclothed ride for much more noble reasons than $50, but when Rose stepped into Ellsworth's major thoroughfare at 5 o'clock in the morning, she was clothed by no more than a six shooter in either hand. The lady had made it well known that she would put a bullet through any eye she saw peeping. Whether Rose's shooting ability was up to carrying out her threat is unknown, but no shots were fired and she got her $50 [5].

The Lady Pirate of Old Sydney Town

Charlotte Badger was a transported felon, possibly bisexual, who lived a fantastic and almost forgotten life. Raised in the slums of London, Charlotte became a notorious pickpocket and was sentenced to transportation to the colonies for life in 1801.

On the voyage to Australia she formed a friendship and possibly a romantic alliance if we read between the lines, with another convict named Kitty Hagerty. Kitty was compliant and placid where Charlotte was bold and audacious and soon allowed herself to be drawn into her friend's plan of escape even before they landed in the fledgling colony at Sydney Cove.

Some time after the pair arrived in the South Land, they were assigned as servants in Van Dieman's Land and they boarded the brig 'Venus' for delivery to the island that would become known as Tasmania.

Charlotte began to exercise her feminine wiles on the ship's mate, Benjamin Kelly, who succumbed to her charms and was soon at the attractive schemer's beck and call. The 'Venus' put into Twofold Bay which would become a whaling centre later in 1828, and Kelly was left in charge while the Captain went ashore. Charlotte worked her magic on the infatuated Kelly and induced him to tap a keg of rum. The Captain arrived back on board to find a Bacchanalian orgy in progress involving Charlotte, Kitty, Benjamin Kelly and another seaman and promptly clapped the quartet in irons.

During the voyage the Captain was forced to release Kelly to help sail the ship but he made it clear he was to be arrested on arrival in Hobart Town.

The Mate stole moments with his beloved when he could and the calculating temptress managed to totally poison his mind against authority, and hatch a daring plan. She drummed it into his head that all he could expect on arrival was a long stretch in the iron gangs. A life of adventure on the high seas with his beloved Charlotte at his side looked far preferable and as they neared Tasmania, Kelly quietly secured the ship's fire arms in the night, and released Charlotte and Kitty as well as some male convicts Miss Badger deemed suitable recruits for a mutiny.

Charlotte took instant control and taking possession of the 'Venus' she ordered the crew and remaining convicts into the ship's boat. The ex-London pickpocket turned buccaneer personally flogged the captain on the deck of the brig before he joined the rest who were cast adrift while Captain Charlotte with the besotted Benjamin Kelly as navigator, set course for New Zealand.

En route, Captain Charlotte ordered the 'Venus' alongside a trading vessel which she and her ruffian crew boarded and took over. Not having the manpower to sail a second prize, they took arms, food, valuables and anything else that took their fancy, and sailed off to land in the Bay of Islands in New Zealand's North Island.

The group was befriended by local Maoris, and Charlotte, Kitty and Benjamin went to live with the local tribes. The rest of the crew sailed along the New Zealand coast raiding several villages to kidnap Maori women. Apparently their last foray ashore went badly wrong and the bold buccaneers finished up as guests of honour and the main course at a Maori feast [6].

Same-sex alliances were apparently not frowned on by Maoris but Kitty is supposed to have moved south to the Rotorua area, where she became the wife of a chief. That is one account, but another has her dying soon after arriving. Charlotte settled down with Benjamin Kelly in what is now Northland.

Several years went by until a British Man-o-War put into the Bay and after hearing of the whites living nearby officers of the ship managed to bribe the Maori tribesmen into revealing their whereabouts.

Sly Charlotte must have got wind of the eminent arrest for she was nowhere to be found, but Benjamin Kelly was clapped in irons. The mutineer was transported to England and hanged. A sad end for Kelly whose biggest mistake was falling head over heels in love with the sea-witch, Charlotte Badger.

In 1818 an American whaler named the *'Lafayette'* bound for Newcastle, Australia from San Francisco, called at the Vavau Islands in the Tongan group to take on fresh water. While supervising the filling and loading of barrels, the whaling ship's skipper was amazed to be addressed in English by someone amongst the watching Polynesians. A white woman clad in native garments stepped from the crowd with a half-caste child at her side. It was Charlotte Badger. She had lived for several years as the wife of one of the Tongans and the child was from their union [7].

By all accounts, Charlotte had not lost her ability to charm sailors and when the *'Lafayette's'* skipper sailed away, Charlotte sailed out of the history books with him at the side of another besotted seaman.

The Eulo Queen

Undoubtedly the best-known female character in the early days of Queensland, was a striking and talented red head named Isobel Robinson. The girl who would become known as The Eulo Queen, was born on the French Colony of Mauritius off the coast of Madagascar in the Indian Ocean about 1850.

She was born Isobel Richardson, illegitimate daughter of James Richardson an army captain, and Priscilla Wright or possibly White. Captain Richardson tried to conceal the evidence of his out of wedlock fling, but by the time Isobel was in her early teens, her parentage was well known. James Richardson packed her off to Australia hoping the memory of his indiscretion would fade when the pert redhead was out of sight and out of mind.

Soon after her arrival in Sydney, Isobel married a man named McIntosh and it appears he died soon afterwards. Isobel travelled to Queensland's central west where she exchanged vows with station manager, William Robinson of *Spring Grove,* Surat, at Roma on 2nd March 1871. William had been born at Richmond in New South Wales and had followed pastoral work until his marriage to Isobel.

It is not known exactly when the couple became involved in the liquor trade, but they were in possession of an hotel in Eulo, Western Queensland by 1886. Apparently a child who died at an early age, was born to the couple. William devoted himself to caring for the horses and outside work while Isobel ran the pub.

Isobel Robinson was described as a voluptuous, auburn-haired beauty inclined to plumpness with a truly magnificent bosom that she loved to show off in décolletage dresses, and adorn with jewellery, particularly opals.

She had been educated in England and spoke fluent French and German. Her ability to charm men of all ages was legendary and beneath a sparkling personality outwardly displayed, she was endowed with an icy business acumen and an ability to effortlessly handle the wildest frontiersmen.

Miners from the surrounding opal fields, shearers, drovers and stockmen all fell for her charm and looks. She held court every night behind her bar extravagantly jewelled and gowned with her red hair piled high, its natural highlights accentuated by sparkling jewelled combs and satin ribbons.

Her forte in the entertainment area was off-colour stories and risqué repartée and the simple men of the outback were enchanted by her looks and wit, staying in her company as long as their money lasted [8].

One story credited to the Queen concerns a barely literate miner who paid his account at her store and kind of lost track of his carefully counted figures after being given 10% discount. The simple soul could not work out this business of paying less than he thought he should have, so he crossed to the hotel where Isobel was entertaining a large group of male admirers as usual. The Eulo Queen was usually known as Maggie to her customers and the confused man pushed up to the bar waving his receipt. "See here, Maggie," he began. "If I was to give you fifty quid how much would you take off?"

The Eulo Queen was too good an entertainer to let a chance like that slip past. She turned to the puzzled miner with a dazzling smile and reduced her bar patrons to howls of knee-slapping mirth by replying "Everything but my earrings, darling. Everything but my earrings."

She was reputed to be an accomplished billiard player, a fine shot with a rifle and extremely gifted at the card table. A little too gifted some claimed and it was whispered on occasion that the pasteboards Isobel played with were marked.

Isobel purchased the Royal Mail Hotel and Billiard Parlour in 1889. By 1900 she owned the Victoria Hotel and in 1902 added The Metropolitan Hotel to her Eulo interests as well as a store and a butcher's shop. Isobel managed to have all her hotel licences cancelled for infringements of the Liquor Act. One of the things that excited interest from the Licensing Board was the fact Isobel was running a private bar in her bedroom. I would venture to say there could be many reasons for such an arrangement. Being suddenly unlicensed was of small concern to the lady now being called The Eulo Queen. She put employees in as dummies and continued trading as before.

On 10th October 1902 William Robinson died aged 63 and was buried in Cunnamulla. Isobel did not appear to be particularly heartbroken and a little over a year later, she wed a Tasmanian named Herbert Victor Grey aged 29 at Eulo on 31st October 1903. Victor Grey was reputed to be Isobel's one true love, but he was also reputed to have a wife in Tasmania at the time of his marriage to The Eulo Queen. It seems likely young Victor was not quite as enamoured as his considerably older bride and also likely that Isobel fell victim to the fatal vanity of women when she gave her age on the marriage document as thirty five. It may have been an honest mistake, however and it is possible that she simply put the numbers in the wrong sequence when she should have written fifty-three [9].

Isobel left Victor in charge of things in Eulo during 1913 and made a trip to England, having made a lot of money out of opals some showered on her by infatuated miners, and others accumulated by a lot of very shrewd gem-buying on her own behalf. She returned as World War I was beginning, and Victor enlisted separating the couple once more.

Some reports say Victor Grey was killed in action, others have him returning injured and dying soon after, but it seems the outbreak of the Great War, followed by the loss of her favourite husband, was the turning point for Isobel's fortunes.

Over the war years, all three of the Eulo Queen's Hotels burnt down and she was left with only her store as a means of existence. What happened to the butchery is not recorded. The Eulo Queen lost her fortune and looks in later years and became overweight and decidedly eccentric. She succumbed to an increasing dependence on alcohol and following a suicide attempt in 1922, Isobel (Maggie) Richardson/ McIntosh/Robinson/Grey left the town she had once unofficially ruled, forever.

The Eulo Queen died at Willowburn, a Mental Hospital in Toowoomba, aged seventy-nine in 1929.

Australia's Annie Oakley: The Claudie Lakeland Story

The exploits of Phoebe Anne Oakley Mozee (Annie Oakley) 1860-1926, born in a backwoods cabin and known as 'Little Sure Shot' have been recounted through song, story, cinema and stage [10]. Annie Oakley was one of the world's great shots with rifle, revolver and shotgun once shooting 4,772 glass balls with a 22 calibre rifle from 5,000 tossed in the air in the course of a day.

Superb shooting by any standards but Oakley's glass ball marathon included 228 misses whereas Claudie Lakeland, a shy teenager who lived deep in the Cape York jungle, was described by her mother Esther, as "a freak who did not know how to miss".

The saga of 'Battling Billy' Lakeland, his wife Esther and their sharpshooting daughter Claudie, had its beginnings when Billy was born in the Sydney suburb of Rushcutters Bay in 1847. William Lakeland was the second child of James and Elizabeth Lakeland nee Anderson.

James Lakeland 1810-1888 was a west Yorkshireman who arrived at Sydney as a convict aboard the *'Captain Cook'* in 1836. He was granted his Certificate of Freedom on 3rd August 1843 and is believed to have taken up farming in the Wellington area of New South Wales [11]. James married Margaret McDonald in 1840. Four children were born of the union – Elizabeth, William, John and Mary.

Just when Billy Lakeland moved north is not known, but in those turbulent times of gold rushes and pastoral expansion, many young native born Australians found it slow on the farm. Billy must have joined those adventurous souls who sought 'Eldorado' because when the Gympie Goldfield opened in 1867, he was there aged twenty.

Billy moved north to the Palmer River Rush on Cape York in 1873 or 1874 and soon became recognised as one of that hardy band of 'gun' prospectors who ranged the wild lands and whose ranks included such notables as James Venture Mulligan and John Dickie.

Billy Lakeland was involved in some very significant prospecting 'tours' as they were known then. In 1874 together with Robert William 'Bob' Sefton, he discovered gold in the Batavia River area and that discovery led to the later establishment of the Batavia Field. Lakeland and Sefton also found gold at Sefton Creek and tin in the Pascoe River country on that trip.

In 1875 Billy is said to have teamed up with one of the north's most colourful characters, Christie Palmerston, and the pair won a fair haul of alluvial gold on the Palmer River.

Bob Sefton put a fifteen-man party together in 1876 with the idea of prospecting on the northern part of Cape York Peninsula. Fifteen well armed and mounted men left Maytown on the Palmer, and Cooktown, in two parties in August. They met by arrangement at the Laura River and worked their way north to arrive at the area that would become known as Coen. Here Sefton split his party into three five-man teams. Sefton and his four men stayed in the immediate area to discover the gold that would eventually found the town of Coen.

Billy Lakeland headed another of the parties and while camped at Croll Creek, he and his men were attacked in the pre-dawn, by Aboriginal warriors. Billy was rolling his swag when a shower of spears flew out of the dawn light. He took two hits in the neck and arm and one of his mates, Jim Watson, killed two raiders and wounded another before the attackers were driven off.

The great bushman's cure-all, OP Rum, was used to disinfect Billy inside and out and after the spear shafts were burned off with heated knives, the party began the long ride to Cooktown to have the barbed spearheads removed.

Sefton and his mates also had their problems with hostile natives and were forced to fortify their camp with earth and rock barricades. Meanwhile, the other party of five which included two Scotsmen, MacDonald and Stewart, were attacked by a war party while travelling, that they described as "hundreds of blacks".

The two Scots fell in the attackers first rush, speared and brained with nulla-nullas. Harry Stuckey, a superb horseman who was in the lead, galloped through a rain of spears to his two companions, Goodenough and Goodfellow. They turned the pack and loose horses back and fled all the way back to Sefton's fortified camp where they alerted the other party.

Bob Sefton, Sam Verge, John Doyle, Jack Russell, Jack Watson, Harry Stuckey, Henry Goodenough and Dick Goodfellow dug in and fought off the hostiles who suffered heavy losses to the prospectors' rifles. After the battle, they returned for the bodies of MacDonald and Stewart. All that remained was bloodstained grass. Some speared horses had been removed as well, so there would doubtlessly be feasting in the Aborigines' camp that night as well as mourning the slain.

Billy Lakeland became a 'name' on the Palmer. He was a cattleman holding *Comet Downs* with his mate Bob Sefton as well as running a butchering business and a brewery in Cooktown in partnership with Joseph Smith Oddy. He married Esther Margaret Culton on 2nd March 1889 at Cooktown. Their first children, a girl and a boy, were born to the couple there. Billy's heart was in the search for the elusive yellow metal however, and he spent long periods prospecting.

In November of 1893 he arrived in Cooktown claiming to be well satisfied with his latest find and intent on bringing his wife and young daughter to live at the claim.

In an interview with a Cooktown newspaper, Billy described his find as being seventy three kilometres north-east of Coen in dense 'scrub' which was the term used to describe the Cape York jungles. Some ladies of Cooktown were horrified at Billy's intention to take a wife and child to such an hostile environment, but Esther was a timber cutter's daughter well used to remote country. Their son Iris Stanley William had died in infancy earlier that year and perhaps the Lakelands felt that tragedy could occur as easily in town as in the Rocky River jungles.

Whatever their reasoning, Billy, Esther and four year old Claudie Ethel May, born 23rd December 1889, set up housekeeping in a neat, pine-log cabin which would be their home for many years.

The Rocky River Scrubs in the 1890s could hardly be considered a site for a normal childhood. A mining warden described it as "being in the heart of the scrub so dense and tall that one is in perpetual shade the whole day".

There were hostile Aborigines as well as more friendly ones whose children became Claudie's sole playmates. There was the whole gamut of tropical problems – snakes in the house, wild pigs in the garden, leeches, ticks, mosquitoes, sandflies, the ever-present spectre of fever and the awful aching loneliness.

Billy Lakeland did well at the Rocky River but like most finds, the hordes of miners who rushed the area, soon cleaned up the easily won alluvial gold and left. Billy prospected for and soon found the reefs but finding them was one thing. Getting the gold out was a horse of another colour.

Billy had a stamp mill shipped to Port Stewart, which was a rather grand name for a tin shed on the fringe of the coastal mangroves north of Cape Melville. With a team of bullocks and a group of Aboriginal helpers, Billy hauled the mill over the Chester Range using ropes made from Lawyer Vines to help the bullocks pull the unwieldy contraption over the worst spots. After a superhuman effort from all concerned, they got the stamper to its site but as hauling a boiler and engine into a region like that was not an option and probably beyond Billy's resources anyway, bush ingenuity had to come into play. A carpenter was imported from Cooktown and soon a waterwheel was constructed and the mill operational. The Lakelands were in business.

By the time she was seven, Claudie Lakeland was her mother's and father's right hand. Billy was often absent mining or prospecting and Claudie helped in mine, house and garden as well as caring for her baby brother, Leo Percy Bruno, who arrived on 23rd February 1897. Esther soon passed the job of official pot-hunter to her daughter. It was in the shadowy half-light of tropical rainforest that Claudie Lakeland honed her shooting skills. A glimpse of a fleeting animal between the towering tree trunks or the flash of feathers from a swooping bird, was all Australia's Annie Oakley needed. Her mother and father were both crack shots but there was something almost mystical about the way hand, eye and weapon came together when Claudie's rifle flew to her bony young shoulder.

All the Lakelands were fine horsepersons and from the time she was little, Claudie had made the trip for supplies perched at first on her mother's knee as Esther rode side saddle, and later riding astride on her own horse. Esther Lakeland was one of the few white women who rode through the notorious Hells Gates on the Cooktown-Palmer River track where so many met their end from Aboriginal ambush.

As the years rolled by Claudie's exceptional skill with rifle and handgun became well known in the north. One day in 1902 or 3 when the blue eyed, pre-teen had ridden into Coen for supplies, she was challenged to a shooting match by local policeman, Roland Walter Garraway.

Claudie shyly accepted Garraway's challenge and in the wide, dusty, main street of Coen, the long-haired, slip of a girl from the Rocky River jungle, wiped the floor with the embarrassed policeman. Had it been in America, every round the combatants fired would have been recorded for posterity, but Australian's penchant in those times for not big noting themselves, has buried the Cohen shooting match.

Roly Garraway had been a Sub-Inspector in the Queensland Native Mounted Police. He was well liked, by white people at least, a superb bushman and more importantly to this story, a crack shot. Had this shooting match occurred in America, Claudie Lakeland would have been an overnight sensation courted by newspaper reporters and Wild West Show promoters. Because it happened on the Australian frontier, it was local news for a while as bushmen told the tale of how "that Lakeland kid could shoot," then like so many of our marvellous true stories, it faded into obscurity so much so that we are not even sure that Roly Garraway was the policeman involved. Some reports say it was the Sergeant at Coen, but Constable Garraway is the most likely choice.

In his own reminiscences while publican of the Grand Hotel at Mackay, Queensland, Garraway referred to the loser as the 'Gallant officer at Coen' and said the highlight of the contest was Claudie's dazzling display shooting pennies thrown in the air with a rifle and with never a miss.

We can only guess how many shots were fired by each, whether they shot with both rifle and pistol, over what ranges, and what the final tally was. What we do know is that the policeman who was no mean marksman, was humbled to the extent that all memories of the event state Claudie blitzed him so badly it was barely a contest. She won and by a long shot at that.

The Lakeland Family
Taken at Mossman 11th October 1900.
(Left to right) Leo Percy Bruno, Esther Margaret, Claudie Ethel May and William 'Battling Billy'. Claudie is looking somewhat disgruntled doubtlessly due to being done up in fancy clothes. It was the first time she had worn a dress.

Photo by Henry J.F. Stains
Courtesy Jim McJannett

Claudie Lakeland as a young woman in 1908. (Possibly a wedding photo).

Courtesy Jim McJannett

The Lakelands stayed on at their home at Neville Creek in the Rocky River Scrubs even after water flooded their mine and forced its abandonment in 1900. Claudie left to marry in 1908. Billy and Esther moved to Coen some time after 1910.

Billy Lakeland stayed in the north prospecting and Sandalwood cutting. He died in late 1921 at Hull River near Coen while going back to a gold find he had made the year before. Family and friends pleaded with Billy not to go out again as his health was bad. Against all advice, he went anyway in late August of 1921. He became ill on the track and turned back. His illness forced him to camp and he died near the head of the Hull River. It was about five weeks before Billy Lakeland's body was found. His remains had been scattered by birds, pigs and dingos. They were collected and buried at the prospector's last camp.

Years later in the late 1930s, John McDonald Hardcastle of the *Cooktown Independent* newspaper, was the driving force behind purchasing, hauling in and erecting a handsome marble monument over Billy Lakeland's grave.

Following Billy's death, Esther lived at Cooktown where she died on 28th November 1949 aged eighty-one. On 21st August 1908 Claudie Lakeland married Percival Thomas Hodges at nineteen years of age. Their first child Myrtle Lucinda was born in Queensland in 1909 and the family moved to Sydney where four more children were born. Norman, Eric, Hazel and Leo R.

Claudie lived out her life in Sydney. We can only wonder if the people she rubbed shoulders with in the streets, parks and shopping emporiums of that cosmopolitan city had any inkling that this well dressed, city matron once humbled the pride of the Queensland Police Force in a dusty, frontier street on Cape York Peninsula.

Claudie Hodges nee Lakeland, died at Balmain, Sydney in 1963.

Lola Montez

There was never a frontier femme fatale to generate more adulation, mystery and scandal than Lola Montez who titillated men and shocked puritans with her lascivious Spider Dance in both America and Australia.

Lola was born in 1818 in County Limerick, Ireland and was actually a dark haired, Irish colleen rather than the mysterious Spaniard portrayed in her stage persona. Born Maria Dolores Eliza Gilbert, she learned to speak Spanish and dance in London and gave her first public performance there in 1844 [12]. She appeared in Europe playing in dramas, comedies and burlesques in Warsaw, Dresden, Berlin, St.Petersberg, Munich and Paris before setting sail for America.

While in Germany, Lola conducted a much-publicised affair with King Ludwig of Bavaria and would later use the liaison as a basis for a stage act.

Lola came to San Francisco in the early 1850s when the fabulous California gold rush was in full swing. People flocked in their hundreds to catch performances of her notorious Spider Dance where she swung to centre stage on a rope that was ostensibly part of a giant web. Lola appeared to be entangled in the web and as she searched her brief chemise and petticoats, she would discover spiders which had to be shaken loose giving her mostly male audiences tantalizing glimpses of her body.

Spider after spider was discovered and the dancing became wilder and more abandoned as Lola hurled the fake arachnoids made of rubber, cork and whalebone from her and stamped them to death on the stage as the music and dance drew to a crashing, whirling crescendo [13]. Audiences loved her as much for her offstage antics as her fiery performances. No respector of truth in advertising, Lola publicly claimed to be the illegitimate daughter of Lord Byron, the poet, and gave newspaper interviews about her numerous husbands and lovers.

She loved to walk the streets of San Francisco smoking cigars with a parrot on her shoulder and two greyhounds on leashes creating a fuss by entering saloons and gambling dens where women were forbidden.

Lola was the toast of San Francisco until a rival dancer Caroline Chapman began performing a parody of the Spider Dance. Chapman's performances brought howls of laughter from the audience followed by groans of desire from the males as Caroline's displays of flesh became even bolder and more salacious than the performances of Lola Montez.

In a fit of pique Lola abandoned the stage to her rival and embarked on a tour of the Californian mining camps. Despite her outward worldliness, Lola Montez hated ridicule and she had trouble dealing with the rough audiences she encountered in the mining districts.

She decided to rest for a while and rented a cottage in a spot called Grass Valley. Miners in the area were thrilled to have a celebrity in their midst and named a local peak Mt. Lola in her honour. A local minister denounced her during a Sunday sermon but Lola got even by knocking on his door in stage costume and performing selected parts of the Spider Dance on his front porch in full public view. She is also supposed to have thrashed a newspaper editor whom she felt had demeaned her in print, with a riding crop.

She obtained a black bear which she kept chained to a stump in her garden. The bear seemed to be more of an accessory than a beloved pet. Lola was reportedly terrified of it and would throw the animal its food from well outside the reach of its chain.

Another local, a miner named Johnnie Southwick who had been fortunate on the gold fields, fell under Lola's spell. Montez was happy to be paid court to, at least as long as the money lasted, and Southwick is reputed to have returned to his claim a poorer but wiser man when Lola tired of retirement and sailed for Australia in 1855[14].

For her first appearance Down Under, Lola Montez chose a burlesque that provided a comic treatment of her affair with King Ludwig called "Lola Montez in Bavaria". Her scandalous behaviour and many lovers who included writer Alexandre Dumas (Snr) and composer Franz Liszt, as well as the Bavarian King, ensured huge interest in her Sydney opening and the event promoters further swelled their profits by selling selected tickets at auction.

Lola's Bavarian comedy was popular in Sydney but when she began performing the Spider Dance it caused an uproar resulting in packed houses and vilification by upright citizens and church elders. The *Sydney Morning Herald* described her act as '*The most libertinish and indelicate performance that could be given on the public stage*' which certainly boosted sales.

From Sydney Lola swept into Melbourne and more sell-out performances as well as mudslinging from the wowsers. A show scheduled for Geelong in Victoria failed to eventuate when she was banned from performing which only made her next audience in Ballarat keener to feast their eyes on this exotic beauty. Lola Montez's greatest Australian success came when she opened in the newly built Victoria Theatre in the gold mining town. Her lacklustre tour of the Californian mining camps had at least given her an inkling of how to relate to the gold-seeking fraternity and the simple diggers of Ballarat went wild, showering her with gold nuggets thrown on the stage.

The editor of the *Ballarat Times*, Henry Seekamp, was a bit more straight-laced than the miners however, and he wrote that Lola's immoral behaviour fostered '*unhealthy excitement*'. Lola responded by drawing public attention to Seekamp's fondness for the bottle and while the editor was pursuing his 'fondness' in the bar of the United States Hotel, she staged an Aussie version of her alleged Grass Valley assault by publicly horsewhipping him. An unseemly brawl resulted with Seekamp throwing a few punches himself, but the newspaperman was tossed out of the pub amid jeers and catcalls for daring to defend himself against the diggers' sweetheart. Lola filed charges against Seekamp that were later dismissed [15].

After three years in Australia, the tempestuous actress and dancer sailed back to California aboard the *'Jane E.Faulkenberg'* and in keeping with the notoriety that constantly surrounded her, Lola's current lover, Noel Follin, was lost overboard during the passage.

Lola gave a few farewell performances in San Francisco then moved to New York. She gave a lecture tour telling her life story with suitable embellishments and wrote a book with the imposing title *The Arts of Beauty or Secrets of a Lady's Toilet with Hints to Gentlemen on the Arts of Fascination.*

The turbulent years seemed to come to an end then and Lola Montez finished her short eventful life working in New York's Magdalen Asylum for destitute women. She passed away in 1861 aged a mere 42 years.

"L'Enchanteresse Espagnole (Lola Montez)" Lithograph by J.D. Middleton, engraved by G.Zobel.

Courtesy The Australian Ballet Collection, The Arts Centre, Performing Arts Collection, Melbourne

Mrs. Black Jack Reed of Booroloola

Henrietta Reed may well have been a member of that sisterhood of good women who marry bad men. The man who dragged her to the furtherest reaches of the ever advancing, Australian frontier was 'Black Jack' or 'Maori Jack' Reed who is remembered as the first settler at Booroloola on the Macarthur River and as a member of 'The Ragged Thirteen' (chapter 4 volume 1.)

Augustus Lucanus, police officer, publican, store keeper, miner, mail contractor and one of the north's earliest citizens, described Reed as "one of the greatest scoundrels unhanged" [16]. Reed's record of fraud, theft, murder and debauchery leaves little doubt to the veracity of Lucanus' claims.

Contrary to reports of a New Zealand heritage, John Ward Reed was more likely to have been born in America around 1836 [17]. The nickname 'Maori Jack' was attached to him but he also carried the handles 'Yankee' and 'Mexican' Reed and was of a swarthy appearance. The first reference to him in Australia is as one of the hardy band who helped policeman Wentworth D'Arcy Uhr bury the victims of Burketown's fever epidemic in May of 1866 (chapter 3 volume 1.)

During 1873 and 1874 Reed was engaged in packing on the Palmer River Goldfield where he also operated a shanty/store/butchery. The first reference to 'Black Jack' and Henrietta as a couple, comes when they were reported running the Beaconsfield Hotel at the settlement of the same name on the Hodgkinson Goldfield.

Between his appearance at Burketown and his time on the Palmer, it seems likely that Reed was engaged in blackbirding in the Pacific Islands. He has been named as first mate for the notorious Bully Hayes. Tales from his time of labour recruiting for the Queensland and Fijian plantations, which was a nice description for slaving, label Reed as a black-hearted villain. One of his less savoury practices was to grab an available native girl after landing at an island and take her aboard ship. After Reed had finished raping the unfortunate woman, he would causally toss her overboard. Sometimes he knocked his victims on the head first. Others he amused himself with by using them for target practice or simply abandoned them to the ocean and the sharks.

Somehow Henrietta came to marry this loathsome man and in his company she became probably the first white woman to cross the north of Australia by the Gulf Track. The Reeds arrived in Pine Creek, Northern Territory in a covered wagon with an Aboriginal helper and nineteen horses on 5th January 1881 having left Normanton, Queensland some months previously.

It would appear Henrietta was well and truly disenchanted with 'Black Jack' before their overland odyssey. A northern Irishman named James McLelland had met the Reeds at Beaconsfield and she was obviously enamoured by the Irish charmer. He and Henrietta made plans to run away together but Henrietta may or may not have been aware that McLelland financed their dash for freedom by helping himself to

£250 in banknotes, two gold rings and a gold nugget breast pin the property of none other than 'Black Jack' Reed [18].

Poor Henrietta. She seemed to have either a penchant or a liking for ratbags as partners and when the bench warrant issued by the Cooktown Court caught up with her lover, she was returned to the tender mercies of her husband.

Henrietta Reed was described in the police description circulated for the capture of James McLelland, as about thirty five years old, five feet five inches tall and rather stout. The battling bush lady must have showed the fortitude of the wonderful women who pioneered this land alongside their husbands good and bad. She endured the trek across the Gulf Track with its wild Aborigines, shark and crocodile infested rivers, rugged terrain and white outlaws to successfully arrive at Pine Creek.

Alfred Searcy, a customs collector who would later feature in the Reed saga, described her as "a splendid woman and as good a bush-woman as one would wish to meet". He went on to say she had often tried to leave Reed but he would follow her like a sleuthhound [19].

'Black Jack' Reed continued to engage in criminal activity during his time at Pine Creek and other new settlements in the Northern Territory. He was before the Palmerston (Darwin) Court on 2nd July and 17th August 1881 for matters relating to possession of a stolen cheque and dishonoured cheques. On 13th December of that year an arrest warrant was issued for him when he failed to appear on a larceny charge [20].

Reed decided the way to make money in the frontier Top End, was to get back into hotel and store keeping as he had done on the Palmer. With this in mind, he and Henrietta travelled back to the east almost certainly by ship, where they acquired a seagoing vessel, the ketch *Good Intent*. The boat was loaded with grog, stores and private loadings for station owners, and sailed from Thursday Island for the western side of the Gulf of Carpentaria.

They sailed up the Macarthur River to the crossing at the future site of Booroloola, which 'Black Jack' had decided would be the place to get rich from the miners, adventurers and drovers moving west. There he made ready to advertise his wares to travellers on the Gulf Track, after running the *Good Intent* onto some rocks only visible at low tide that have been called Reed Rocks ever since.

The area was occupied by a motley crew of horse and cattle thieves who claimed to live trapping brumbys from mobs that had bred up in the Limmen River country following a disastrous droving trip by Thomas Price Cox who lost all of a mob of thoroughbreds he was bringing across the Gulf Track in 1873 [21]. These exceptionally fine animals were the nucleus of the Limmen wild horses that were joined by other animals lost by drovers, prospectors and teamsters. They at least gave a legitimate excuse for being in the area to men like 'The Orphan' Jack Martin who had served five years in Queensland for robbing Chinese travellers [22], and Jack Sherringham who was wanted for horse stealing in Queensland [23], and some other shady characters.

The Reeds met an enthusiastic bunch of potential customers on the banks of the Macarthur where they threw up a rough shanty. They became particularly popular when 'Black Jack' and his first mate Harold Best, a large Negro native of North Carolina whom the locals promptly christened 'Smoked Beef', began unloading their stock of grog which included a large amount of square, green bottles labelled 'Come Hither' [24]. The stuff certainly brought this uncivilised crew's crazy instincts hither and a wild party began on the banks of the river. C.E. Gaunt a non-drinker who witnessed the revel and later wrote about it, claimed *"anyone with a few drams of 'Come Hither' under his belt would charge hell with a bucket of water"*.

'The Orphan' got involved in a gunfight with a man called 'One Eyed Billy' and managed to blow his own thumb off. A Queensland detective who was seeking Jack Sherringham, was lured from his quest and after a session on 'Come Hither' jumped into the river and was taken by a crocodile in full view of other partygoers. A man named Dick Morris attempted to take his leave with a skinful and drowned himself and his riding and packhorses, trying to swim the Macarthur.

We can assume Henrietta mainly concerned herself with serving drinks and collecting money as 'Black Jack' socialised with his customers. Tongue loosened by grog, Reed bragged to someone that on the last leg of his journey up the river, he would have put a bullet in his first mate if he got the chance to save paying his wages. Reed's remark got back to Harold Best now known as 'Smoked Beef' and the big American Negro tore into 'Black Jack' giving him an awful thrashing and practically demolishing the hastily built, tin and bush timber shanty [25].

Reed left Henrietta to run the Macarthur store and shanty feeling she would have a hard time running from the remotest settlement in Australia, and sailed down river again to ascend the Roper River where he planned to deliver station loading and probably do some more sly grogging at Leichhardt's Bar.

It was at Leichhardt's soon to be known as Roper Bar, that 'Black Jack' and the *'Good Intent'* struck a rough patch. On 29th January 1885 Sub Collector of Customs, Alfred Searcy, was inconsiderate enough to arrive at Leichhardt's Bar in the launch from the ship *'Palmerston'* which was anchored downriver. He promptly arrested Reed and impounded the *'Good Intent'* and its contents for unpaid duties [26].

Searcy then proceeded overland to the Macarthur River and found Henrietta conducting business there. As there was only an iron and timber store and such goods as it contained, Searcy impounded that and to be sure of the duty owed, he also took possession of Henrietta's personal belongings as well. It was at this time that Henrietta Reed's best-known exploit occurred.

With tears in her eyes Henrietta pleaded with the Customs Officer to allow her to keep an old concertina of little monetary but great sentimental value. Searcy was not immune to dewy-eyed, feminine pleading and he handed the treasured squeezebox back to the lady. It was lucky Alfred Searcy did not attempt to coax some notes from the instrument when he held it, because the concertina was stuffed with over a thousand pounds worth of cheques [27].

On 9th March 1885 'Black Jack' and Henrietta Reed were convicted of illegally shipping dutiable goods and falsifying clearances and fined £50 [28]. The *'Good Intent'* was towed to Port Darwin where it was sold by V.L.Solomon and Co. to the Northern Territory Lightering Company for £225. on 28th March 1885. This amount together with the value of the confiscated goods reached £500. It was the period after the seizure and sale of the *'Good Intent'* that 'Black Jack' Reed became associated with the Ragged Thirteen (chapter 4 volume 1.) The *'Good Intent'* was wrecked between Darwin and Port Charles in 1892 [29].

Not content to leave his dream of a pub at Booroloola unachieved, 'Black Jack' applied for a licence for the proposed Royal Hotel at Macarthur River on 8th April 1886. The application was granted two months later. Reed went in for carting as he had done previously, and Henrietta ran the pub until it was sold to William Taylor in March 1887 [30].

The Reeds stayed in the Top End with 'Black Jack' escaping from the Burrundie lockup while incarcerated for stealing a horse named *Normanby* on 25th February 1888 during his time as one of the Ragged Thirteen.

It seems that by this time 'Black Jack' may not have been as possessive of Henrietta as in previous years. He toddled off to play the bold adventurer with his larrikin mates in the Ragged Thirteen. That bunch of hard-cases must surely have their popular status as fun-loving larrikins questioned by their association with a man like Reed. But to be fair to them, they may not have been aware of the vile deeds of his blackbirding days. From the time of leaving Booroloola, 'Black Jack' and Henrietta appear to have led more separate lives.

Henrietta may have left Port Darwin for Sydney aboard the *'S.S.Menmuir'* in April of 1892 but by May of 1893 she was definitely the licencee of the Wayside House at Marble Bar, Western Australia [31].

'Black Jack' was convicted of sly grogging at the Star of the East Mine at Nannine, Western Australia on 25th August 1893. Three weeks later, he was again charged with sly grogging at Yahagoug. He was in court again in 1895 for sly grogging and stealing at Mt.Magnet [32]. In 1895 he served six months jail for robbing the till at the Metropole Hotel at Murchison and was operating a store at Nunyarra near Eucalyptus by 1899.

'Black Jack' Reed eventually moved to New Guinea where his death was recorded on 3rd July 1907. Reed's occupation and place of residence at his passing, was given as miner of Misama Island, Papua [33]. In an interview with an old seaman in the 1970s, historian Jim McJannett heard the old man speak of Reed as a trader not a miner, and name him as 'Yankee' Reed who operated a ketch named *'Smuggler'*. It looks as if he had returned to his habits of blackbirding days and the old salt told Jim, Reed had grabbed a girl on a beach. As he was dragging her towards his boat, the girl's brother ran into a store run by a Chinese, where he grabbed a gun and raced back to the beach shooting Reed dead in the surf. If this admittedly unproven story were true, it would be no more than black hearted 'Black Jack' deserved.

Henrietta Reed could certainly be regarded as a fitting mate for 'Black Jack' on the available evidence. She shot through with another bloke. She obviously had no objection to selling grog, sly or legally, and she put it over Alfred Searcy like a seasoned con artist. The thing that suggests she was a good woman who stuck to her husband for better or worse and perhaps picked up a few of his tricks along the way, was the amount of positive comment about her from people like Searcy and Lucanus who were present at the time.

Little more solid evidence has been heard of Henrietta following her keeping the pub at Marble Bar, but Jim McJannett has collected other information which he says may or may not be right. Another old sailor and waterfront worker from Australia's north spoke of a Henrietta marrying again after John Ward Reed died at Misama Island. The story continued that she retired to Cairns and lived in a neat cottage with a brass wall plate bearing the name 'Misama'. At least this tale, fact or fiction, gives the story a happy ending. The only question remaining is, did Henrietta name her cottage out of respect after the final resting place of a man she loved unreservedly and supported actively in his shady dealings, or as a celebration of the place that finally ridded her of 'Black Jack' Reed.

Cockney Fan

Fannie Haynes was a London Cockney who became a storekeeper and publican in the Northern Territory. She reputedly has an interesting connection with the American West.

Fannie came overland from Charters Towers, Queensland in the 1890s with her husband, a man named Cody, and they began a store on a goldfield named Wandi. Her husband was reputed to be a relation, possibly a brother, of William Frederick Cody known throughout the American West and later the world, as Buffalo Bill, Scout, Indian fighter, buffalo hunter and showman. (chapter 7 volume 1).

Cody died and Fanny went on to marry Tom Crush, a member of Parliament with whom she established the Federation Hotel at Brocks Creek in The Territory. 'Cockney Fan' as she was known, seems to have been a bit hard on husbands because Tom Crush died as well and she then married Harry Haynes, a one legged man. Perhaps he was easy to catch.

A well known, Territory anecdote about this down-to-earth lady concerns the time a cyclone came tearing down on Brocks Creek shortly after a travelling priest had arrived. Fannie and the Priest were standing under the pub's veranda as the wild storm came tearing in uprooting trees and ripping off roofs [34].

"Quick" cried the man of God. "We'll pray."

"No we bloody well won't" replied the pragmatic lady publican. "We'll cover up the bloody flour!"

The Lady Bushranger

Australia has seen a female outlaw or two like Elizabeth Jessie Hunt who travelled with buckjump shows under the alias Miss Kemp as a roughrider, whip cracker, trick shot and rope spinner. She was born at Carcoar, New South Wales to James and Susanne Hunt in 1890, and apprenticed to a circus and buckjump show run by Martin Breheny who was an accomplished acrobat commonly known as Martini. Breheny had travelled with 'Lance Skuthorpe's Wild Australia' and later formed his own show going on to own the great buckjumpers *Dargin's Grey* and *Bobs*. (See chapter 14).

Elizabeth, commonly known by her second name, Jessie, became the Australian Ladies Roughriding Champion in 1905 and 1906 and was said to have been the child bride of her mentor Martini who died when she was seventeen. There is no record of a marriage to Martini and Breheny left his assets to his father after he was killed at Armidale, NSW when his horses ran over him after being spooked by a train in 1907.

Elizabeth Jessie Hunt ran 'Martinis Rough Riders' until 1911 when she suffered an accident and arranged the sale of the show to Alf Neave, a West Australian who amalgamated it with Phillip Lytton's 'Australian Buckjumpers' and toured England under the banner of 'Wild Australia'.

Jessie appears to have fallen on hard times then. In August 1912 she was tried for theft and served some time in a reformatory. Shortly after that in 1913, she was again arrested using the name Jessie McIntyre and spent over a year in a Sydney Women's Prison.

She seems to have managed to stay out of trouble until 1920 when she married Major Ben Hickman at Sofala, New South Wales, but Hickman cleared out to India with a shipment of horses for the Army Remount Trade.

Jessie gave birth to a son and because she was destitute and unable to care for him, she gave the child up for adoption. At some time Ben Hickman turned up again and the two attempted a reconciliation. Things did not work out for the second time and Jessie fed up with life in general and men in particular, packed up a few horses and fled deep into the Wollemi Ranges, part of the Great Dividing Range in central New South Wales [35].

During the 1920s Jessie Hickman became one of the leaders of a gang of cattle duffers who moved stolen stock through secret tracks in the rugged country between Singleton, Mudgee and Portland. She lived in a hidden cave in the ranges and proved too elusive for police, staying in the cattle thieving business until she was eventually captured on 2nd May 1928.

Jessie had a few near misses before the 'traps' eventually yarded the woman who became known as 'The Lady Bushranger' although she was never a 'bail up' artist. In one incident, she reportedly leapt a horse off a cliff top ten metres into a river and swam out of sight of a policeman and black tracker.

Jessie Hickman faced Magistrate Mr. H.H.Farrington on the 9th May 1928 at Rylstone Court House where she was committed for trial at the Mudgee Quarter Sessions 28th August on a charge of stealing six head of cattle [36]. The case apparently hinged on one easily identifiable cow and just before the trial the animal in question disappeared and was found dead and burnt in the bush. Some of Jessie's cattle-duffing mates were not about to see the Lady Bushranger do any more time and in a scenario repeated many times in Australia, the jury declared Jessie Hickman 'not guilty' of cattle duffing.

Jessie returned to the Wollemi Ranges and lived there until her death in 1936 from a brain tumour caused by the accident that forced her to stop running Martini's Buckjump Show.

Jessie Hickman may have had a fairly cavalier attitude to other people's stock, but the Lady Bushranger was greatly admired for her spirit and self-reliance.

America's Frontier Ladies

Dora Hand, Diva of Dodge City

To this day, Dora Hand remains one of Dodge City's most mysterious memories. Not cast in the same mould as other 'Nymphs de Prairie' or 'Calico Cats' who frequented Dodge and other Kansas trail towns, Dora became a symbol of all that was right and pure in the hell-raising cow town. Some researchers claim her life has been glorified but the story of Dora Hand contains large doses of the romance and adventure of the old West.

Her real name was unknown and she performed under the stage name of Fannie Keenan. Apart from vague rumours of a genteel Boston upbringing and an education in music and the arts in Germany, only one man in Dodge City knew her history and he took the knowledge to his grave [37].

James H.Kelly fought in some Indian Wars with Colonel George Armstrong Custer and was given the nickname 'Hound' Kelly in Hays City because of the pack of racing greyhounds he kept. He moved to Dodge and with Peter L.Beatty, Kelly opened the Alhambra Saloon and Gambling House. The partners also operated the Varieties Theatre as well as a Dance Hall.

Somehow the 'Hound' tag was changed to 'Dog' and Kelly became the first Mayor of the Queen of the Cow Towns known to one and all as Dog Kelly.

Dora Hand may have met Kelly in one of the other Kansas cattle towns because she had sung and danced at Abilene and Hays City before she became established as the protégée of Dodge's Mayor.

Kelly set Dora up in his house which they shared with another of his dancehall girls named Fannie Garretson. Dora's dainty form and air of wistful innocence won the hearts of the rough westerners who flocked to the Varieties Theatre to watch her nightly performances. As she swayed to the music and carolled old favourites and lilting ballads, men of the sordid trail towns were briefly transported to a finer and more genteel way of life. Dodge City was simply and uncomplicatedly in love with Dora Hand.

Dora was one of the few people admired on both sides of the tracks that bisected the cow town. Somehow she managed to bridge the social divide and was popular with the moneyed, solid citizens of Dodge City who often invited her to sing at functions, as well as the rowdier element who frequented Whisky Row and the Red Light area.

Obviously cultured, she could have moved in eastern society circles and the scandal or wild streak in her make-up that led her to the harsh Kansas border regions will never be known. Men fought for a smile and a kind word from the Toast of Dodge and when a young Texan called Jim 'Spike' Kenedy pressed his suit a bit too ardently, Dog Kelly's bouncers tossed the young cowboy into the street.

That is one version. Another claims that Kenedy quarrelled with Kelly over a fine for carrying a gun and a disorderly conduct charge. Spike Kenedy felt himself 'above the law' and indeed he was no commonplace cowhand. He was the son of Mifflin Kenedy, rancher and riverboat operator who set up the legendary King Ranch with Richard King. (chapter 5 volume 1). Kenedy and King divided their holdings in 1868 and Captain Mifflin Kenedy's share became the huge 'Rancho de Los Laureles' of Nueces County, Texas.

'Spike' Kenedy was a darkly handsome young hellion who had already been in brushes with the law but his father's money and position had managed to smooth things over. Swearing vengeance on Dog Kelly, the Texan retired to brood on his grievances and in the early hours of 4th October 1878, he fired four shots through the wall of the house where Dora lived believing Kelly to be sleeping there. One of Kenedy's bullets smashed Fannie Garretson's bedpost and another passed through a flimsy partition killing Dora Hand instantly where she lay in Kelly's bed alone.

The description of the group which pursued Kenedy given by the *Dodge City Times*, was '*as intrepid a posse as ever pulled a trigger*', and the claim was no lie. It was a showcase of legendary Western Lawmen led by Sheriff Bat Masterson with members Assistant Marshall Wyatt Earp (chapter 2 and chapter 2 volume 1), Marshall Sam Bassett and Deputy Sheriffs Bill Duffy and Bill Tilghman who appears in chapter 3 volume 1.

For two days the lawmen ran Kenedy's tracks south west of Dodge until a heavy storm wiped all signs away. They obtained information from a ranch that the fugitive had passed and picked up his trail only to lose it again in another storm. The posse members were unsaddled, resting their horses and allowing them to graze, when a horseman came into view. Bat Masterton recognised the horse and the way its rider sat the saddle. It was none other than their quarry. The men stayed hunkered down with their gear as the Texan drew closer until at about seventy metres, Spike Kenedy recognised members of the posse and went for his gun.

Wyatt Earp threw his rifle to his shoulder and killed Kenedy's horse with one shot as Bat Masterson drove a bullet through the murderer's shoulder. Kenedy fell with his horse and also sustained broken bones when the animal crashed to earth on top of him.

They hired a wagon and team to haul the wounded outlaw back to Dodge City where he lay in bed until he was well enough to face trial. Mifflin Kenedy travelled from Texas to be at his son's side and when Judge R.G.Cook held a preliminary hearing in closed quarters at the tiny Sheriff's Office, he released Dora Hand's killer for lack of evidence. Wyatt Earp later claimed it was Mifflin Kenedy's money that prevented his son facing a jury but karma played its part and 'Spike' Kenedy failed to totally recover from his wounds dying in Texas a few years later [38].

The funeral of Dora Hand was one of the biggest events seen in the cow town to date. No Boot Hill burial for the Darling of Dodge. Her expensive casket, the finest the town could supply, led a huge crowd of mourners northwest to the new Prairie Grove Cemetery. Wealthy ranchers and affluent businessmen rubbed shoulders with gamblers and gunmen. Respectable matrons marched beside dance hall girls and whores as Dodge City from its leaders to its dregs, laid their sweetheart to rest.

As the text for his service the officiating minister chose *'Let he who is without sin cast the first stone.'* [39]

Many years later as he lay dying in the Fort Dodge Old Soldiers' Home, James H.'Dog' Kelly steadfastly refused to reveal the details of the former life of Dora Hand/Fannie Keenan to any of those who came seeking the knowledge. Kelly died some time in 1912 and the secrets of Dodge City's mysterious lady died with him.

Belle Starr, The Bandit Queen?

America managed to produce quite a few lady outlaws with titles like 'Rose of Cimarron', Pearl Hart, 'Cattle Annie' and 'Little Britches' but like her male counterpart Jesse James, it is Belle Starr who wears the crown as Queen of the female desperadoes.

There is little actual evidence to suggest Belle took part in anything except a few very small scale crimes, but she showed her flair for the illegal in organization and management.

Born Myra Belle Shirley at Carthage in Jasper County, Missouri on 5th February 1848 Belle's parents were well-to-do farming people. The Shirleys sent their daughter to the Carthage Female Academy where Myra Belle learned her three Rs as well as piano playing and other skills calculated to enhance her opportunity of marrying well.

Horses and guns were the particular passion of young Myra however, and when the Civil War began raging across pro-slavery Missouri and abolitionist Kansas, she was already a daring horsewoman and an accomplished shot with rifle and revolver.

Myra Belle Shirley was fifteen when guerrilla leader William Clarke Quantrill and his raiders sacked Lawrence, Kansas. Men like Jesse and Frank James and the Younger brothers rode with Quantrill and Myra idolised the grim bearded horsemen. She was soon operating as an informant for the rebel guerrillas in spite of objections from her parents and when the Shirley home was burned in a Union raid on Carthage, John Shirley moved his family to Scyene, Texas.

Myra's father and mother placed her in school again hoping she would forget her wild ways, but in 1866 Cole Younger rode into her life following his first bank robbery in company with his siblings Jim and Bob and the James brothers. The gang whose exploits would be told all around the world, rode back to Missouri in 1867 leaving Myra Belle a legacy from handsome Cole Younger in the form of a thickening waistline.

John and Eliza Shirley were mortified by their daughter's scandal, but being publicly ostracised didn't seem to bother Myra Belle. She left the child, a girl she named Pearl Younger, with her mother and flitted off to Dallas where she danced, sang and dealt card games in saloons for a while until she took up with another of the outlaws who would fascinate her throughout her short life [40].

Jim Reed was a horse thief born in Missouri and operating around the Dallas area. Much to her father's displeasure, Reed married Myra and took her and Pearl back to his parents' place at Rich Hill, Missouri. Jim Reed was one of a large group of outlaws loosely allied to Texan John Fischer who moved stolen horses north for sale. It was with Reed that Myra got her first taste of organised crime. She loved it and never again would the wild child of poor John and Eliza Shirley be found far from illegal activity.

Jim Reed had to flee after shooting two brothers named Shannon, and the family lived in California for a time where Myra's second child, Ed Reed, was born in Los Angeles. A policeman learned of Jim's identity and the family fled again, Myra Belle returning to her family in Texas while Jim Reed hid out in the Indian Nations.

Myra Belle regularly visited her fugitive husband at Sam Starr's Ranch. Starr was a Cherokee and one of the most notorious Indians in the Nations, constantly at loggerheads with the tribal councils. His property was a hangout for all sorts of undesirables, just the sort Myra Belle Reed liked.

Jim Reed came to grief on 28th April 1874 when a stage hold-up went wrong and he was pursued and shot by Deputy Sheriff John T. Morris of Collin County, Texas. His apparently not very grief-stricken widow farmed her two children out and moved in with the gang full time. She did not take an active part in stock theft, whiskey-peddling, burglary or armed robbery, but she planned the jobs and seemed to have a genius for securing quick releases through the Court system for any members taken by the law. She also bestowed her favours on several gang members before settling on old Sam's son as her next consort.

Myra married Sam Starr Jnr. and being the wife of a Cherokee allowed her to claim tribal land of her own. She established herself on the Canadian River naming her holding 'Younger's Bend', probably after her first love Cole Younger. On 15th February 1883 Belle Starr as she was then known, and her husband faced Judge Isaac Parker's court at Fort Smith, Arkansas charged with horse stealing. The pair were sentenced to a year's jail each and served nine months before returning to Younger's Bend.

Sam Starr was soon hiding out in the woods again following a robbery of the Creek Nation Treasury and with Sam separated from her once again, Belle's ardour for the Indian outlaw began to cool.

Belle tried to elope with a new recruit called Jim Middleton who was wanted for robbery, arson and murder but after the pair split up to avoid being seen together, Middleton failed to turn up at the arranged rendezvous and was found dead from a shotgun blast. Sam Starr was the obvious suspect but he needed to stay hidden and soon Belle was pursuing new adventures with a white criminal known only as 'Blue Duck'.

Blue Duck went on a drunken rampage killing a farmer and Belle managed to get his death sentence commuted to life imprisonment and a year later to somehow obtain a pardon.

It was during this time that a photo was taken of Belle and her latest beau which clearly shows Belle Starr was no great beauty. Perhaps her outlaw support network was more appealing to residents of Younger's Bend than the lady herself. As a young woman she sported an attractive figure and was reputedly well versed in the arts of seduction. Whatever the reason, Belle Starr was obviously attractive to men and a well known figure in the lawless Indian Nations as well as being a thorn in the side of Isaac Parker, the Hanging Judge. (chapter 3 volume 1).

Following his release Blue Duck only flew briefly to his

Belle Starr stands with her handcuffed lover, a white outlaw known only as Blue Duck.

Photo: Archives & Manuscripts Division of the Oklahoma Historical Society. Photo no. 4631 Fred S. Barde Collection

lover's arms before being shot by 'persons unknown' in July 1886. Once again the obvious suspect was Sam Starr.

Belle was saved the considerable effort and expense of burying any more lovers hunted down by her inconsiderately, jealous husband when Sam Starr got into an argument with a tribal enemy named Frank West. Starr and West reached for their weapons at a dance at Whitefield in Indian Territory and both men managed to kill each other on 18th December 1886.

There was never a shortage of drama in Belle Starr's life. In April 1887 her illegitimate daughter Pearl Younger produced an equally illegitimate granddaughter. Belle had the nerve to be scandalised sending Pearl away with the child to spare embarrassment to her 'good' name. Her son Ed Reed was sentenced to seven years for horse stealing on 22nd July 1888 but Belle went to work on his behalf getting him pardoned within a few months. Obviously horse theft was much more acceptable than unwed pregnancies at Younger's Bend.

Between rushing to her son's defence and spurning her daughter Belle managed to conduct liaisons with Jack Spaniard, a Chocktaw Indian, and Jim French, a Creek, before being wed once again. This time the bridegroom was a tall young Creek Indian named Jim July who was under indictment for horse stealing when he took up residence at Belle's place.

On 2nd February 1889 a few days before July was due to face Judge Parker's court, he set out for Fort Smith to answer the charges. Belle rode along intending to accompany her husband as far as San Bois about twenty four kilometres away. On her return trip Belle stopped at the home of a man named Rowe to visit his wife and while there she met and quarrelled with Edgar A. Watson about some farmland he had been trying to lease from her.

About 3 o'clock that afternoon Milo Hoyt and another man were at the ferry over the Canadian River when they heard shots and saw Belle Starr's white horse run down the road and swim the river. Hoyt went back up the trail to find Belle lying dead from two shotgun wounds [41]. Jim July obtained a continuation of his case and came back to lay a charge of murder against Watson. The charge was dismissed in Judge Parker's court for lack of evidence.

Evidence surfaced later that suggested Jim July may well have circled back to kill his wife and then made a fast trip to Fort Smith to secure an alibi, and there were also whispers that Belle's son Ed Reed may have been the culprit.

Belle Starr was never the Bandit Queen popular history has portrayed her as but she was an imposing figure who always rode side saddle, dressed in elaborate outfits and seemed to be a magnet for trouble. It was her fascination with outlawry and her rather loose ways for the times that brought her in contact with the owlhoot brigade, and her ability as an organiser and bush lawyer that secured her place in the annals of Western Crime.

Stagecoach Whips, Pony Express Riders, Bullockys and Mule Skinners

Swift scramble up the sidling where teams climb inch by inch;
Pause, bird-like, on the summit – then breakneck down the pinch;
By clear, range-country rivers, and gaps where tracks run high,
Where waits the lonely horseman, clear cut against the sky;
Past haunted half way houses - where convicts made the bricks -
Scrub-yards and new bark shanties, we dash with five and six;
Through stringy-bark and blue-gum, and box and pine we go -
A hundred miles shall see tonight the lights of Cobb and Co.!.

Henry Lawson

Similar Beginnings

Australia and America shared a lot of common ground in the various modes of transport through their respective wild wests. Early Australian coaching ventures were largely unsuccessful until they embraced America's hard won knowledge of vehicle design suitable for rough, bush tracks and long distances.

Iron sprung, British style coaches offered a comfortable ride by the standard of the times, but would only perform satisfactorily on made roads, which were in very short supply in early Australia. Efforts made to traverse bush tracks in the new country were met by impossible delays as blacksmith forged, iron springs broke with monotonous regularity. The common conveyances for settlers were unsprung drays that crashed over logs and rocks with such bone-shaking ferocity that walking alongside was a far preferable alternative to riding.

The leather-sprung, thoroughbrace coaches designed in the United States of America, and known as 'Concords' solved the problems of rough, Australian roads and it was four experienced, American stagecoach men who founded Cobb and Co and pioneered long distance coaching in Australia.

Remote districts in both countries utilised packhorse mail, and sometimes freight services. Bullock and horse teams were the motive power for shifting goods and settlers across the two continents by wagon. Steamboats utilised the inland waterways and coastal waters of both countries. Railways made much quicker progress in North America with most areas having reasonable access to a railhead thirty years after the Civil War. Australia on the other hand is still only marginally serviced by rail in its remote areas.

Australia's Early Days

Cobb and Co – Australia's Wells Fargo

An idea conceived by a young American from Massachusetts named Freeman Cobb and put into effect by him and three other Americans, gave birth to one of the greatest transport companies on earth [1].

The partners imported Concord coaches, harness and horses from America and when the first leather-sprung, thoroughbrace coach on Australian soil pulled out of Collins Street, Melbourne on 30th January 1854 to run to the Forest Creek diggings in half the time of previous trips, a coaching empire that would crisscross Australia and spread to three other countries, was born.

Freeman Cobb and his partners John Lamber, James Swanton and John Murray Peck, were quick to capitalise on the knowledge they had gained working for American coaching firms The Overland Mail Company, and Adams and Co. Cobb and Co soon outstripped its rivals by establishing horse changes every twenty kilometres or so and greatly increasing the speed of service, to become the leading coach line to the Victorian gold fields and beyond.

The original partners sold their interests in the firm in 1856. Cobb and Co was destined to change hands several times until being purchased by Canadian, Alexander Robertson and John F.Britton in 1860. Britton did not stay long in the partnership his share being taken over by another Canadian, John Wagner.

Robertson and Wagner continued to use the name Cobb and Co and were soon joined in the venture by William B.Bradley, Walter R.Hall, William F.Whitney and James Rutherford – all fellow North Americans. James Rutherford was the major motivating force behind the expansion of the company. He managed Cobb and Co for fifty years and was responsible for forging it into the biggest coaching concern in the world.

James Rutherford was a man of huge vision with the guts and energy to back it up. No desk manager, he travelled constantly keeping tabs on the far-flung enterprise. Rutherford's favourite means of touring the lines and relay stations was a cape-cart or pole sulky. A sulky was normally fitted with shafts and used as a conveyance for one or two people pulled by one horse. Fitted with a pole and harnessed to two of the big raking 'coachers' of mixed draught and thoroughbred blood favoured by Cobb and Co, Rutherford could achieve exceptional daily distances on his inspection tours.

With a ready supply of horses at his disposal, James Rutherford regularly averaged around 200 kilometres per day for days on end. One of his pole sulkies is on display at the Cobb and Co. Museum in Toowoomba, Queensland and an interesting fact of its construction is the way the swingle-bars and traces are attached with solid leather straps rather than the usual metal links. This would have made the vehicle travel very quietly without the rattle of chains and metal fittings, and no doubt it allowed Rutherford to surprise tardy employees by not announcing his approach.

Some years later, Sir Sidney Kidman also utilised pole sulkies to travel his chain of stations and the 'Cattle King' averaged similar distances to Rutherford. Light weight, tall wheels combined with two big, free going horses made pole sulkies very light to pull and consequently easy on the animals.

Under Rutherford's guidance, Cobb and Co spread into every State of Australia except Tasmania and coaches bearing their logo also ran in New Zealand, Japan and South Africa [2]. At the height of their operation in Australia during the 1880s, Cobb and Co coaches travelled over established routes spanning six states, harnessing 10,000 horses daily. They had coach and harness making factories and workshops in Bathurst, Goulburn and Hay in New South Wales, Castlemaine in Victoria, and Brisbane and Charleville in Queensland catering to outside orders as well as their own needs.

The company had extensive pastoral properties breeding horses, sheep and cattle as well as general stores and dealerships. This coupled with their overseas interests made Cobb and Co the largest coaching firm on earth.

The twentieth century saw a gradual decline as motor transport began to take over. James Rutherford died on 13th September 1911 and the company battled on adding motor trucks and shedding horses and coaches until Fred Thompson drove the last horse-drawn service from Yuleba to Surat in Queensland on 14th August 1924.

The Great Depression sounded the death knell for the firm that had dominated Australian mail and passenger delivery for seventy years. On 30th June 1929 Cobb and Co went into voluntary liquidation at a meeting in Brisbane signalling the end of an era in Australia's history that will never be forgotten.

Australian Master Reinsmen

Handling the 'ribbons' on multiple hitches of between four and eight horses some only partly broken-in, was no job for an amateur. To hold a position as a stagecoach driver, a person needed to be a talented horse handler and also able to cope with the various dramas involving passengers and goods in transit.

Stagecoach drivers commanded a great deal of respect in frontier communities and some individuals have been long remembered for their skill at the reins.

Coach driving seemed to be an occupation that attracted extroverts and those types that would be known today as 'people persons' [3]. Joe Hirschberg, a North

Queensland Cobb and Co driver, had a booming voice, an extrovert's manner, and was a great socialiser as well as a dancer and billiard player par excellence.

'Ointment' Taylor of the New England region always delivered a slick sales pitch to his passengers with the object of selling a few jars of his home made cure-all salve.

Bob White of the Warrego region would only stop relating yarns and anecdotes when someone managed to get a word in edgeways with a joke he had never heard. Bob would then hang on every word and gleefully re-tell the tale at the first opportunity.

In marked contrast was 'Silent Bob' Bates who once ignored a travelling salesman's comments about a fine, wheat crop beside the road between Glen Innes and Inverell in New South Wales. The coach reached the same point on the return journey and the salesman who was still a passenger was surprised to hear Bob remark out of the blue "It's oats, ya fool."

Victorian driver Mick Dougherty was renowned for his tall stories and was once entertaining a lady passenger with a long winded yarn of how he had trained a kangaroo to collect a mailbag from the coach then sort and deliver the letters. The coach rounded a turn to find a big, old man 'boomer' right beside the road. Dougherty gave the 'roo a casual wave, called "Nothing today, Jack!" whereupon the 'roo hopped into the scrub and the lady turned incredulously towards the poker-faced reinsman who launched into his next story without so much as a smile.

Australia's best-known Cobb and Co driver was a Tasmanian named Edward Devine who bore the nickname 'Cabbage Tree Ned' because of the cabbage tree hat he habitually wore.

Ned Devine began driving coaches between Geelong and Ballarat in 1854 and soon acquired a reputation for easily handling large teams of horses. Cobb and Co experimented with a huge vehicle in 1862, capable of carrying around seventy passengers on the Geelong/Ballarat route. The massive coach was named 'The Leviathan' and must have been a striking sight harnessed to twenty-two matched greys with Cabbage Tree Ned at the reins. The Leviathan proved too unwieldy for long term service and its use was discontinued, but Ned's expertise with the huge team has become one of coaching's treasured memories.

When the first English Cricket team to tour Australia arrived in 1862 they were met by their official transport, Ned Devine driving a brand new coach with a twelve-horse team supplied by Cobb and Co as their contribution to the touring players. Cabbage Tree drove the coach with the cricketers on board, right onto the playing field at Geelong to start their match in spectacular style and at the conclusion of the tour, Ned Devine was presented with a handsome award of 300 sovereigns by the grateful Englishmen [4].

Cabbage Tree Ned must have been looking for a change of scenery by 1863 when he moved to New Zealand, but he stuck to his profession and employers, driving coaches both north and south from Dunedin in the South Island for Cobb and Co.

Ned's time in the Shaky Isles contained some epic feats of driving and a lot of aggravation for the humourless Scots who sub-contracted mail routes from Cobb and Co. in the Otago Province. The larrikin Aussie was constantly in trouble for failing to attend to unimportant details (to him) like filling out waybills and also for playing practical jokes on influential passengers.

In atrocious conditions on the Pigroot Track, which went from Palmerston near the east coast, to the Central Otago Goldfields, Ned had a coach overturn. He was travelling at a walking pace around a hillside between Palmerston then called Shag Valley, and Dunback then called Waihemo, when the nearside wheel horse slipped into its running mate on the bottom side due to treacherous footing in the muddy conditions that gave the 'Pigroot' its name.

Ned saw the coach was going over but he urged his horses to take the strain long enough for passengers to scramble clear as the vehicle balanced precariously on the slope with its topside wheels off the ground. The driver made a leap for the high side but his coat caught on the luggage rack and Ned went down with the toppling coach [5]. Luckily he was thrown from harm's way and in a short space of time, the vehicle was righted and the journey resumed. This incident happened about two kilometres from the farm where I grew up, but a long way before my time.

Ned Devine at the reins of a six-horse hitch at Palmerston, Otago, N.Z. during the 1870s. The author lived near Palmerston as a boy.

Photo courtesy Otago Settlers Museum, Dunedin, New Zealand

Ned was taken off the 'Pigroot' run when he played one joke too many, but a very old lady whose family ran the Pigroot Creek change station during coaching days and who used to visit my family when I was young, remembered her mother telling her of a lovely coach driver named Mr. Devine who always had sweets in his pockets for little children [6].

Ned was transferred to the northern run but the jokes didn't stop. He once turned a passenger's stomach after the man bought some fish from locals at a seaside town named Hampden on the north run between Dunedin and Oamaru. Ned's passenger was very proud of his lovely fresh purchase but the coach driver seriously remarked "What a pity it was that this particular species was poisonous at this time of year." Divine launched into a straight-faced discourse about ocean currents carrying polluted materials into the feeding grounds due to prevailing weather patterns and the final upshot was the disillusioned traveller leaving the fish on the coach seat on arrival at Oamaru, and Cabbage Tree Ned having a tasty supper.

In 1878 Ned Devine hung up his whip and retired. Little is known of his later life but he was admitted to the Ballarat Benevolent Asylum on 12th July 1904 where he passed away four years later on 18th December 1908. In 1937 Cobb and Co's Old Drivers Association of Victoria was instrumental in having Ned's remains moved to a prominent part of the Ballarat cemetery and erecting a handsome monument over his grave.

The grave of 'Cabbage Tree' Ned Devine, Cobb and Co. Master Reinsman

Photo courtesy Neil McArthur

Australia's Pony Express

The short and glorious history of America's Pony Express is known around the world, and is recounted later in this chapter, but few people are aware that Australia also had an organised Pony Express Service of even shorter duration.

The Australian pony express is closely tied to construction of the Overland Telegraph from Darwin in the Northern Territory to Port Augusta in South Australia, which began in September 1870. The telegraph line that would span the continent from north to south, was to join with an undersea cable laid from Java connecting Australia with the world [7].

South Australia had been chosen over Queensland as the State to carry the line, and the construction deadline was 1st January 1872. Failure to complete by the due date would render the South Australian Government liable for a £140 per day penalty. Due to delays mostly caused by the northern tropical monsoon season, the line was not completed on time. By June 1872 the situation was desperate. There were still over 450 kilometres of overland line to complete and the State Government hatched a bold plan to ensure the incomplete service could function.

John Lewis was a South Australian cattleman who had leased a pastoral run on the Coburg Peninsula east of Port Darwin. Together with his brother James and two mates called Walter Soward and J.Peterson, John left Adelaide on 18th January 1872 to travel their horse plant overland to the new property.

Not far to the south of Barrow Creek Telegraph Station, Lewis and his companions had a clash with warlike tribesmen who fired the grass attempting to cut the travellers off from reaching the station, but the four hard riding bushmen pounded through the flames and galloped to safety at the new, heavily fortified telegraph building.

It was at Barrow Creek Station that Lewis was offered the chance for him and his horses to become Australia's Pony Express Service.

John Lewis and his men pushed north to Tennant Creek where the horsemen were contracted to start their key to key relays between there and *Daly Waters* on 1st June. They established a halfway point for north/south changeovers, which changed as the distance shrank, and the pony express swung into action.

John Lewis and his brother Jim, kept twenty-five of their best horses and sent Soward and Peterson on north with the rest of the plant. They engaged a man named Ray Boucaut to help run the relay service, and for nearly three months, the riders moved back and forth along the shrinking distance between the telegraph keys. The trips were not made without risk as the men were in constant danger of attack from hostile natives particularly the warlike Warramunga who had attacked explorer John MacDouall Stuart's party and had slain drover John Milner a year before at Attack Creek which was right on the express route.

For safety's sake the men travelled in pairs and it was necessary to camp and spell the horses. John Lewis and his men had no means of setting up horse changes along the track of their hastily organised venture, and the same horses made the whole trip over either the north or south sections.

The South Australian Government's agreement with Lewis specified three men but in his memoirs he mentions a fourth. In the warlike regions they travelled, the only practical way would be for the men to take turns sleeping and standing guard [8].

The Lewis brothers once found that hostile Aborigines were not the only danger to their health when they were nearly shot by telegraph construction workers. The constant threat of attack by aborigines had made the O.T. workers a bit trigger-happy and when Lewis and his off-sider were coming into one of their campsites after dark they were met by a volley of rifle and revolver shots. The yells of surprise from the Express riders and more than likely a few easily recognisable Anglo-Saxon expletives, soon proved the approach to the camp was not being made by hostiles and luckily no one was hurt in the exchange.

A few weeks after the service started, the battling construction teams got a respite when the Java/Darwin cable suddenly broke down on 24th June. From then on, the Express riders carried north-bound messages only.

The north and south telegraph line builders drew gradually closer together as they sunk holes, stood poles and ran wire. The Pony Express riders galloped over the ever-shortening miles keeping the incomplete, marvel of technology operational with pounding hooves and dripping sweat until finally on Thursday 22nd August 1872 north crews met south and the wires were spliced in thick lancewood scrub 644 kilometres south of Darwin. The sea cable was repaired by October putting Australia in instant Morse Code contact with overseas.

John Lewis, his brother Jim, Ray Boucaut and the fourth man Albert Hands, had set up and run Australia's Pony Express Service. It is doubtful if the enterprise was ever given an official title and it certainly was nothing like the slick operation set up by Russell Majors and Waddell in America.

John Lewis, Grazier and operator of Australia's Pony Express

Photo courtesy National Library of Australia

Lewis and his galloping dispatch riders were simply bushmen who happened to be on the spot with a horse plant when a means of connecting the keys of the embattled Overland Telegraph was desperately needed. The hastily organised and short-lived Australian Pony Express Service was a tribute to the adaptability and courage of our pioneers and at least it didn't go broke like it's American counterpart. One thing we can be sure of, is that none of the four Australian riders would have been unduly upset when the telegraph put them out of business.

Freighting on Two Continents

Freighting goods in America seems to have been largely in the hands of big operators. This would no doubt have evolved from the need to travel in large parties as a defence against Indian attack. Immigrant settlers had always banded together in wagon trains for the same reason, and although Aboriginal resistance was a very real threat to travellers in Australia, war parties were smaller and less mobile than the horse-borne tribes of America.

Most Australian carriers were small operators often owning only the team they drove, but it was common practice for teamsters to travel in convoy not only for protection in early times, but also to assist each other by double teaming their wagons through difficult terrain like steep hills and river crossings.

Both bullocks and horses had their devotees. Bullocks were slower but were much easier to yoke to the wagons than horses that wore a full set of harness. Bullocks were also cheaper to buy, survived well on natural herbage and as they were less likely to wander, did not require hobbling at night.

Some incredible treks were undertaken and completed by slow plodding bullock teams. On 18th March 1895 James Whitmore's team rolled into the town of Eucla[9] on the South Australian/Western Australian border. Whitmore and his twelve bullocks had begun their journey in Queensland and travelled 3,860 kilometres to reach the town on their way to the gold fields of Kalgoorlie and Coolgardie, 650 kilometres ahead of them. Whitmore's was the first bullock team to cross South Australia's Nullarbor Plain.

An effort was also made to use Asian water buffalo as team bullocks in the Northern Territory. The buffalo team left Southport near Darwin and travelled twenty-nine kilometres which was a very satisfactory day's stage under extremely hot conditions. When the team reached Collett's Creek they did what buffalo would consider sensible behaviour and marched straight into a waterhole, wagon and all, to the dismay of the teamster [10].

Horse teams were a lot quicker travellers and Australian teamsters found that by crossing drafts with saddle horses, a nuggetty type of team horse that became known as a 'clumper' was produced. These horses were able to survive on native herbage much better than pure bred draughts and did not require extra feeding.

Large bullock teams hooked up for a celebration in Parkes, NSW in the 1920s. Working bullock teams were usually yoked in pairs like the team at left.

Photo courtesy Ned Winter collection

Bill McArdle's horse team hauling wool near Brewarrina, NSW in early 20th Century. McArdle was a noted roughrider who was drowned swimming a horse across the flooded Barwon River in 1930s.

Photo courtesy Ned Winter collection)

Although used in many areas of Australia, mules never achieved the wholesale popularity that they did in the United States, as draft animals. Donkey Teams were very popular in dryer regions however, because the thrifty little beasts did very well on sparse herbage. Large hitches of forty or more donkeys made a picturesque sight hitched to big loads of supplies and produce. In America the 'Mule Skinner' was popularly known as a rough individual with a remarkably inventive ability in the use of profanity. This crown was worn by the bullocky Down Under and the 'bullock punching' profession produced 'Lords of Language' that have been celebrated in song and story.

The Humble Packhorse

Packhorses, donkeys and mules made a huge contribution to settlement of both continents. Remote mining operations in mountainous regions owe their survival to the patient plodding equines which brought supplies in and ore out. American ranchers and Australian run holders used packs extensively to supply and shift their stock camps and in Australia almost all long distance drovers moved their gear by packhorse rather than horse drawn vehicles so they could follow the grazing without worrying about vehicle access.

Huge pack teams operated in early times due to the rugged nature of the country. Some mines in North Queensland brought all their ore out by pack train to the smelters with some teams containing over 300 animals.

In 1886 Andy Wolfgang left Georgetown in Queensland for the Kimberley gold rush with stores for the miners loaded on 140 packhorses. Wolfgang traversed the Gulf Track and it is a pretty safe bet he and his helpers would have been heartily sick of loading and unloading packs by the time they reached the Kimberleys [11].

Almost anything can be carried by packhorse provided weight limits of around forty-five kilograms per side (ninety kilograms total weight) are observed. Machinery has been taken apart and packed into remote areas. Fencing and building materials, game carcasses and trophies have been carried in and out of the back country by ingenious packers who could work out a way to carry almost anything on their patient charges.

Building material packed into New Zealand's back-country by the author in the 1960s.

Photo by Jack Drake. Courtesy Jack Drake collection

The nomadic herdsmen of both countries utilised a packhorse or two to carry their worldly goods from job to job and the Australian Bagman had his parallel in America's Saddle Tramp.

A uniquely Australian invention was galvanised water canteens bought in pairs and hung on each side of the packsaddle. Canteens carried twenty-two litres each and were imperative in Australian conditions where it often took a day or two to travel between waters.

Packing equipment differed between countries but the basic principle of carefully balanced loads on either side, was much the same. Australian packsaddles had a tree with metal hooks attached and padded by a straw stuffed panel on either side. The American, Sawbuck style pack saddle had the tree bars joined over the top by crossed pieces of timber and instead of coming with its own padding, was placed on top of a heavy saddle pad.

Australian packers placed goods to be carried in large leather or greenhide pack bags fitted with two rings each that hung on the packsaddle hooks and were held firm by a long leather surcingle that passed right around them and the horse. Americans used bags and sometimes boxes known as panniers, and lashed the whole load down to an extra girth under the animal's belly by means of a rope tied in either a 'Squaw' or a 'diamond' hitch to secure the load.

Clyde packed in traditional Australian style at a packing school, 'Red Gum Ridge', Stanthorpe Qld 14.3.1998. Loaded pack bags topped by swag and Bedourie oven. Instructor Jack Drake at left.

Photo courtesy Jack Drake collection)

Packhorse Mailmen

'The mail must go through' has always been the byword of postal services around the world and the more remote parts of Australia and North America presented their own set of problems for postmen.

Although the dashing, Pony Express rider captured the public's attention as the ultimate, horseback postie, a lot more letters were delivered over the years by packhorse mailmen who undertook huge journeys through remote regions on a regular basis.

The Fizzer

Australia's most well known, Outback Mailman was Henry Ventlia Peckham who was immortalised as 'The Fizzer' in Mrs Aenas Gunn's book, *We of the Never Never*. Peckham carried the mail from Katherine to Anthony's Lagoon around the beginning of the twentieth century and he always had a fear of perishing from thirst – a very real possibility in the arid regions he rode through. The 'Fizzer' was on his first trip on the Katherine to *Victoria River Downs* run when he found altogether too much of the water he valued so highly, drowning as he attempted to cross the Victoria River on 11th April 1901 [12]. Peckham's last words shouted to his Aboriginal helper as the water swept him away, were "Save the mail!"

Bureaucrats don't understand Flooded Rivers.

Packhorse Mailman Mick Madrill whose father Ambrose introduced 'broncoing' to the Australian Cattle Industry (chapter 6 volume 1) once arrived at the King River in the Northern Territory on the last leg of his western run into Katherine in the first few years of the twentieth century. The river was flooded with no chance of swimming his horses. Aware that if he failed to get across, station orders he was carrying would miss the train to Darwin and delivery would be delayed for weeks, Mick took drastic action [13].

A rusty wire cable was slung over the river to make a primitive flying fox that had been used in times of need before, so Mick unpacked his horses and after turning the animals loose and covering his packs and bulky items with a tarpaulin. He put his clothes, hat and boots in a waterproof bag, broke the seals of the mailbag and putting the letters in with his clothes, he slung the bag on his back and worked his way hand over hand across the cable. Mick made the crossing with his body dragging in the water and his hands getting cut about by burrs on the old, worn out, wire rope. Madrill dragged himself ashore on the north bank and after donning his clothes he shouldered the bag of letters and walked the thirty odd kilometres into Katherine.

Mick's fellow citizens of the north were all impressed by their mailman's dedication, but alas the powers that be in the postal service saw things very differently. Mick Madrill had broken the seals of His Majesty's Mail, an unpardonable breach of regulations, and instead of being complimented on his devotion to duty, he received a severe reprimand and the threat of prosecution should his heinous crime ever be repeated. You can't win!

'Speargrass' Jack Gard. Gulf Mailman

Jack Gard's father William was a miner who came to the Palmer and Hodgkinson gold fields in the mid 1870s when those rich alluvial and reefing finds were a magnet for hopeful miners from around the globe. William Gard married an English girl from Lancaster named Mary Ellen Holt at the Thornborough Court House on 14th February 1881 and a large brood of young Gards was soon growing up on the North Queensland goldfields.

Young John Joseph Gard, born three days after Christmas in 1897, was number eight of the thirteen children born to Mary Ellen and William and although some of his siblings left the area, the Gulf had claimed Jack and he was to spend his life there. From an early age it became plain Jack was a born bushman and as a young man he was dubbed 'Speargrass' because people reckoned he knew every blade of grass in the Gulf country.

Horses were the love of his life. He worked as a stockman and horsebreaker, trapped and broke in brumbies, ran a few cattle and only took work at his father's profession of mining when he really needed a job.

Speargrass was one of the last of the Packhorse Mailmen. He took the contract for Georgetown and return via *Islands, Huonfels, Ironhurst* and *Dagworth* Stations in May 1944. By 1956 he was getting a bit modern and using an early four-wheel drive Landrover for his mail run but when the monsoon came down it was back to running in the horses and loading the packs.

'Speargrass' Jack Gard takes the sting out of a fresh horse in a Gulf country stockyard.

Photo courtesy Gard family collection

Speargrass had taken a chance in March that year thinking the worst of the wet season was over and was at *Huonfels* Station when Cyclone Agnes tore in off the Coral Sea to rip Cairns apart. Agnes swept across the coastal ranges flooding the Etheridge River and the full force of the water hit *Huonfels* at daylight on 8th March.

The Aplin family who owned *Huonfels* were awakened to rapidly rising water in their house. Speargrass was staying at the homestead and he, Charlie Aplin, his wife and brother George took the only escape and climbed for the roof. George Aplin didn't make it, the waters tearing his grip loose as he tried to scramble onto the corrugated iron. The last words he spoke were a shouted "Goodbye Charlie" to his brother as the floodwaters took him away [14].

A short while later, there was an ominous shudder as the whole house shifted, then it was off its stumps and racing downstream. The building hit a huge gum tree with a rending crash and broke up. The three on the roof made a desperate grab for the branches. Charlie Aplin and Speargrass got a hold but Mrs. Aplin's grip failed and the men watched horrified as she was borne away on a section of roof.

Perils like that don't come the way of the average suburban postie and the two men spent an uncomfortable and grief stricken night in the tree until the water fell enough to reach a nearby ridge where Charlie's daughters who had witnessed their mother and uncle being swept away, waited.

Speargrass was probably glad he was travelling by Landrover that tragic time in March, 1956 or some of his beloved horses may well have gone the way of his friends, the Aplins. The vehicle was lost but that would not have meant much to a horseman.

Jack 'Speargrass' Gard lived his life out in the Gulf country and finally entered the Garden Settlement Retirement Home, Mareeba on the Atherton Tableland. He died on 11th November 1983 and is buried at Mareeba [15].

Days of the Dromedary

One mode of transport used in Australia's early times that was never utilised in America, was the camel. Huge desert and semi desert areas with watering points uncertain and far apart made this country a natural place to introduce camel power.

The first dromedaries in Australia were imported from Peshawar, India with their Sepoy drivers in 1860 to be used by the ill-fated explorers Burke and Wills. When they marched out of Melbourne on 20th August 1860, the camels formed one line and the horses and bullocks another. It takes quite a while to accustom stock particularly horses, to camels. One of the new immigrants was a bit cantankerous and the *Melbourne Herald* reported an amusing incident when a fat policeman was bowled over but the 'wide' arm of the law took it in good part and joined in the general laughter [16].

Burke and Wills didn't exactly thrive in arid regions but the camels did finding Australia very much to their liking. Sir Thomas Elder was the next importer. He brought 120 riding and pack camels in to service his large pastoral properties in South Australia.

Camels remained popular in the Outback until well after World War I when they were superseded by motor transport. Large populations of Afghan drivers formed communities in places like Marree in South Australia and Cloncurry in Queensland. They were strict Moslems and each settlement had its own Mosque [17]. Cameleers were lumped together under the collective title of 'Afghans' in Australia but they came from lots of different middle-eastern regions and many became wealthy and respected citizens.

Camels were extremely strong animals capable of carrying loads of 450 kilograms or more. When packing wool, most of them carried four bales. Australia also made use of them as draught animals and camel teams of up to twenty animals were a common sight. They used simple harness with draughthorse collars around their necks placed upside down because they fitted better that way.

Australia was probably the only country to pull stagecoaches with camel teams. During the West Australian gold rush that began at Coolgardie in 1892, cattle king Sid Kidman and partners were running coaches as well as Cobb and Co. Competition was intense and Cobb's found they could handle desert lines that would be impossible for horses by hitching camels to their thoroughbrace coaches. Camel teams were also used by coach operators in western New South Wales during drought times.

One of the first jobs of the legendary bushman's outfitter, R.M.Williams, was as a camel handler for Billy Wade, a sailor turned missionary, who attempted to spread the gospel among South Australian Aborigines [18].

Camels were left to run wild when motor transport put the teams out of business. In recent years camel racing has become popular and Australia is now exporting them back to their ancestral homeland in Saudi Arabia. Australian feral camels are proving faster than the finely bred racing camels of the Arabs and it is ironic to think that brumby bred Aussies are beating the blue bloods on their own turf or should that be sand?

Old meets new. Camel and horse teams at Duchess Railway Station, Qld

Photo supplied by Cloncurry & District Historical & Museum Soc. Inc. With permission from the Vipen family

Riverboats

Settlement of many inland regions in America and Australia was by water. Both continents had a river system navigable by ocean going vessels for many miles inland, and Riverboats with a shallow draft were soon in service to navigate upper reaches and tributaries.

The first steamboat to ply America's river trade was the *'New Orleans'* built in Pittsburgh, Pennsylvania in 1811. Her maiden voyage was accompanied by an earthquake which must have sent a shiver of foreboding through her operators, but she went on to a successful career as the first paddle steamer of the hundreds that would create the riverboat legend.

The huge floating hotels of the Mississippi with their luxurious saloons and frock coated gamblers are known worldwide, but inland U.S.A. was opened up by trading and freighting vessels that plied the 'Mighty Mississip', the 'Big Muddy' – the Missouri, and other rivers to their limits of navigation. Chicago, Illinois on the shore of Lake Michigan, had a large fleet of steamers plying the Great Lakes which helped build it into one of the world's major trading cities.

The main jumping off points for almost everything that went west by land in America, were St.Louis and Independence on the Missouri where goods and passengers for the west changed over to wagons and coaches. St.Louis was the home of the United States Fur Trade in early times (chapter 1 volume 1) when traders known as *'voyageurs'* brought bales of pelts down the river by keelboat. The Chouteau family who were big operators there, built a special shallow draft steamboat called *'The Chippewa'* in 1859 which navigated the upper reaches of the Missouri thousands of kilometres inland almost to the Rocky Mountains near Fort Benton, Montana [19]. Upper Missouri riverboats competed successfully with overland freighting firms.

Steamboats were one of the most profitable enterprises in early California with local operators banding together to form the California Steam Navigation Co. in 1854. The company held a monopoly over shipping in the San Francisco Bay and upriver to Sacramento and beyond [20].

Riverboat operators on the Columbia along the borders of Oregon and Washington States, developed a relay of steamers connected by portages around rapids at the Cascades and the Dalles, in 1850, that probed the river almost to Walla Walla in the south east of Washington State. Columbia River trade was monopolised as well when riverboatmen John Ainsworth, Jacob Kamm, Simeon Reed and R.R.Thompson pooled their vessels to form the Oregon Steam Navigation Co. [21].

Australia's early inland settlement happened in much the same way with the first navigation of the Murray and Murrumbidgee Rivers happening in 1830 when explorer Charles Sturt travelled them by whaleboat. Sturt went about the first trip on our inland waterways in a rather perverse manner hauling his specially constructed craft overland from Sydney in sections to the upper reaches of the Murrumbidgee then exploring down the river to its confluence with the Murray and the Darling before lack of supplies compelled him to return upriver to his starting point [22].

The riverboat trade began in the early 1850s when the South Australian Government offered rewards of £2,000 for the first two people to operate commercial riverboat services. A young miller named William Randell built the *'Mary Ann'* at Mannum on the lower Murray, South Australia, with the help of a local blacksmith and two bush carpenters. She was an unlovely creation 59 ft (18 metres) long with a homemade engine, but she worked.

Captain Charles Francis Cadell R.N. was the owner of a much more handsome craft. Cadell had a two engine 125 ft (38 m) steamer constructed to his own specifications in Sydney. He steamed around the coast in his new paddle wheeler the *'Lady Augusta'* and set off upriver from Goolwah near the mouth of the Murray in August, 1853 secure in the knowledge he was the first operator in business and one of the prizes was his [23].

As the *'Lady Augusta'* steamed along Cadell suddenly became aware he was not alone on the river. The homely *'Mary Ann'* was at Euston when Cadell's imposing craft swept past. William Randell fired his cumbersome boiler that was chained to the deck and the race was on.

The *'Lady Augusta'* beat the *'Mary Ann'* to Swan Hill. Cadell collected his reward and Randell missed out because his boat did not meet specifications demanded by the government. This began a rivalry between the two captains that lasted until Cadell quit the river in 1859 going back to sea only to be murdered by a mutinous crew in the Pacific Islands.

Riverboats opened up the country all along the Murray, Darling and Murrumbidgee Rivers. Riverside farms prospered many growing into huge stations with imposing homesteads overlooking the river. Echuca on the New South Wales/Victorian border was linked to Melbourne by rail in 1864 giving it the monopoly on the goldfields trade.

Echuca became the St.Louis of Australia and its largest inland port. When the gold petered out, it took over the wool trade and at the height of the riverboat era in the 1870s, 1880s and 1890s the town boomed like few rural centres ever have. Like America, the trade became highly organised and controlled by family companies. William McCulloch and Co. were one of the largest operators with fifty-five boats operating through the 1880s [24].

Flat bottomed paddle wheelers with a very shallow draft towing barges of the same construction, worked their way up the Darling as far as Bourke in north western New South Wales to bring wool down and sometimes drought conditions stranded them in the upper regions. The *'Eliza Jane'* was once trapped by falling water near Bourke so her skipper set up a sawmill to cut Redgum timber which was always in demand for boat hulls. The *'Eliza Jane'* took three years to return to her home port, Goolwah, on that trip but when the floods came she brought a massive load of sawn Redgum down the river for a handsome profit [25].

Australian riverboats plied the big inland waterways until the 1960s and there has recently been a revival in traffic for the tourist industry. Because of the Redgum hulls and solid construction of the old steamers, many hulks were found fit for restoration and once more they thrash their way up and down the 'Mighty Murray'.

The American Transport Experience

John Butterfield and the beginnings of Wells Fargo

The first man to make a big name in America's coaching industry was John Butterfield who won the contract to set up a stagecoach line from St.Louis, Missouri to San Francisco, California in 1857. Butterfield opened the line in partnership with another New Yorker, William G.Fargo, and their first crossing of the west took twenty-four days arriving on the Pacific coast on 10th October 1858.

Butterfield and Fargo's venture known as the Overland Mail Company, ran a reliable service twice a month for the next three years impressing the Post Master General and leading to them being awarded other mail contracts for routes from Kansas City to Stockton, California, and from San Antonio, Texas to San Diego, California [26].

The partners were already active in California where they had reduced rates driving their west coast rivals, Adams and Company, into bankruptcy in 1855 by exactly the same means that their former employees Freeman Cobb and associates used to establish supremacy in Australia. Butterfield and Fargo's dominance of major western routes cemented them as the major operators of the time. Fargo and Butterfield took Henry Wells on board and the legendary transport company of Wells Fargo was born.

Wells Fargo also ventured into banking and following the Civil War they obtained a monopoly over the United States stagecoach business when their two major competitors dropped out. John Butterfield left the firm in 1860 and Wells Fargo went from strength to strength becoming a giant in finance and still active today as American Express.

Russell Majors and Waddell, Freighting Giants, and Coaching 'Wannabees'

With the spread of immigrant settlers across the west and the impact of the California Gold Rush, freighting from the railheads and steamboat terminals in the region of the Mississippi and Missouri Rivers became big business. Alexander Majors, a freighter and trader, stole a march on his competitors by using oxen instead of mules for motive power to his wagons. Oxen could forage more successfully on the great plains and in much the same way as bullock wagons and drays proved their worth in early Australia, their slower pace was more than offset by reliability and pulling power.

By 1855 Majors was in a position to expand further and he took in two partners William H. Russell and W.B. Waddell [27]. Before the next three years was up, the company was running 3,500 wagons and employing some 4,000 men.

William Russell tried to interest his partners in a stagecoach line to the west coast in opposition to The Overland Mail Company, but Majors and Waddell were sceptical of its viability without a government mail contract. Russell went ahead alone and true to his partners' prediction, the venture failed.

Russell was bailed out by his partners who took the struggling stage line on as part of the company. This proved a costly mistake, but the entrepreneurial William Russell waved a light at the end of the tunnel to his worried partners by proposing the establishment of the Pony Express Service.

The Pony Express was one of the most glamorous and romantic ventures in the west. Lightweight riders galloping night and day on relays of fast horses, could cross over half the continent in an average of eight days with one blistering run being completed in just six [28]. Letters were written on light-weight paper and cost $8 each to deliver. Russell Majors and Waddell put the whole ambitious project into action in a mere sixty days and the advertisement they placed for riders would hardly have stood up to today's discrimination laws. It read:

> *WANTED – Young Skinny, wiry fellows not over*
> *Eighteen. Must be expert riders, willing to*
> *risk death daily. Orphans preferred.*

The first run started on 3rd April 1860 with riders leaving St.Joseph, Missouri and Sacramento, California simultaneously on the 3,163 kilometres journey. Forty riders to the east, and forty to the west were in the saddle at all times, and relay station staff were expected to have fresh horses saddled and ready when the riders dashed in. An express rider would throw the leather 'Mochila' with its four locked mail pouches, over the stripped down, lightweight saddles that were used, leap aboard and thunder off on the thirty kilometres odd trip to his next change of horses.

For eighteen glorious months, the romance of the Pony Express captured the imagination of America and the world, but when the Pacific Telegraph Company and the Overland Telegraph Company joined their Trans American line and tapped out the first message on 24th October 1861, the Pony Express was finished, and with it Russell Majors and Waddell.

Ben Holladay 'The Stagecoach King'

For a time during the 1860s, a rough-hewn frontiersman named Ben Holladay became the biggest freighting operator in America with 20,000 wagons, 150,000 draft animals and 15,000 employees. Holladay became known as 'The Stagecoach

King' but passengers on his cross-country runs did not swing along in the leather slung Concord coaches commonly seen in the movies [29].

The Concord manufactured by Abbott Downing and Company of Concord, New Hampshire, which became the model for Australia's famed Cobb and Co coaches, made its first appearance in California in 1850 but up until the mid 1860s, Ben Holladay's passengers travelled on unsprung mule wagons with a body that could be dismantled and used as a primitive barge at river crossings.

Freighting was Holladay's main focus and as far as passengers were concerned, if they wanted to get there, they took things the way Ben and his employees dished them out. This made for uncomfortable days bouncing over the prairies; nights sleeping on the ground, and a rough diet of predominantly salt junk and beans.

Ben Holladay was a 'hands on' manager who often travelled with his wagons. He was an unpolished individual with few social graces, but a huge amount of natural business acumen. Holladay was once described by high-minded, railroad promoter Henry Villard, as illiterate, coarse, pretentious, boastful, false and cunning.

In spite of Villard's damning description, Ben Holladay was a natural born survivor. In 1862 he foreclosed on his competitors Russel, Majors and Waddell whom he had been propping up with financial infusions as he waited for the right time to strike. This came in the wake of the ill-fated Pony Express.

In 1865 as the Civil War began drawing to a close, he made a good deal on Army surplus wagons and animals and further extended his empire, and in 1866 with an eye on the ever encroaching railroads, Holladay sold his whole operation to Wells Fargo at a handsome profit.

The illiterate backwoodsman who had done so well out of animal powered transport, then invested his money in railroads and shipping and presumably lived happily ever after.

An anecdote often told about Ben Holladay relates how a boy from one of the frontier hamlets along his stage route was told the Biblical story of Moses [30]. When the Sunday School teacher reached the part about the children of Israel spending forty years in the 300 mile desert, the lad fixed the storyteller with a disbelieving stare. "Forty years" he cried. "Only 300 miles! Humph! Ben Holladay would have fetched them through in 36 hours!" Probably something of an exaggeration but in the youth's eyes, Ben had Moses shot to bits.

America's Best Whips

Two names stand out among American stagecoach drivers. Charlie Parkhurst and Hank Monk. Both were exceptional rein handlers and both had charismatic personalities that set them apart from their everyday colleagues.

Charlie Parkhurst was a rugged character with one eye who was often called 'the greatest whip in the west'. Parkhurst kept a huge 'quid' of chewing tobacco in his cheek practically all his waking hours and would let fly with a stream of juice at regular intervals with complete disregard for anyone in the immediate vicinity but his dexterous handling of four sets of reins when driving an eight horse team was a sight to behold [31].

While travelling without a guard, Parkhurst was once held up by a bandit who called himself 'Sugar-Foot'. Smarting over the loss of the mailbag, and what Charlie regarded as a slur on his competence, he began practicing with his revolver to prevent the shameful situation happening again.

Some time later Charlie was piloting his team when 'Sugar-Foot' made a second appearance, but there were no easy rewards this time. Parkhurst swiftly transferred the reins to his left hand, put down the whip, and his right flashed to his holster. The stage driver's colt roared and bucked bringing the bandit's career to an abrupt end as he cartwheeled from his horse.

In 1879 the stage driver died, but the undertaker who took charge of the Greatest Whip in the West's body, made a strange discovery. Charlie Parkhurst was a woman!

Hank Monk was another Master Whip or 'Jehu' as they were sometimes known in America, whose rein handling skills were legendary. Unfortunately a fondness for the bottle affected him somewhat in latter years, but in his prime, Hank Monk could turn a running six horse hitch in the space of a Western street and bring them to a stop exactly where he wanted them time after time [32].

Monk's ability with the ribbons was so complete he made the job look easy enough for other drivers to claim he was "the luckiest man to ever climb on top of a box" but luck had little to do with it. Hank Monk was a master.

The most famous story about Hank happened when he transported Horace Greely from Carson City in Nevada to Placerville, California in 1862. Greeley was the editor of *The New York Tribune* and one of the most widely read journalists of his time. Before departure, he informed Monk he had to be in Placerville to deliver a lecture and would appreciate it if Hank could travel with all possible speed [33].

In hindsight Horace Greely came to bitterly regret his instructions as Hank Monk threw the big Concord around the torturous twists and turns of the Sierra Nevada Mountains in a better than fair imitation of a madman. The coach body rocked so hard on its leather springs when they tore along one section of the mountain road, that Horace's head crowned with a badly squashed top hat, splintered through the coach's roof panelling just behind where the grinning Hank sat on the driver's seat.

Regretting his previous concern about haste, Greely began imploring the driver to "Slow down man!" but Monk threw his long whip out to crack over the racing team yelling above the din of hooves and wheels to "Keep your seat Horace. I'll get you there on time!" And he did.

History has not recorded whether Horace Greely was in a fit state to deliver his lecture on arrival at Placerville, but thanks to Hank Monk, he was certainly there with time to spare.

Black Bart – State of the Art Stagecoach Robber

A highwayman or road agent, who made himself very unpopular with Wells Fargo management in California, was a dignified, well spoken robber who called himself Black Bart. He was courteous and humorous, never robbed a passenger and never fired the shotgun he used to menace stage drivers and guards with. Black Bart always appeared on foot on uphill grades where the horses or mules were reduced to walking pace, and wore a flourbag hood over his head.

Between 26th July 1875 and 3rd November 1883 Bart held up twenty-seven Wells Fargo coaches and escaped with thousands of dollars. In his last hold up stage driver Reason McConnell, managed to wound the bandit with a rifle and he fled dropping a handkerchief. A cleaner's mark on the hanky was traced by Wells Fargo detective, J.B.Hume, to a San Francisco laundry the owner of which knew the man it belonged to [34].

Black Bart turned out to be a popular, well dressed character named Charles E.Bolton and he served four and a half years in San Quentin prison for his eight and a half year's living off Wells Fargo.

There is a rumour that refuses to go away in spite of there being no record of it in Wells Fargo files, that the company paid Bart a stipend for life on condition he gave up robbing them [35].

Bart was of a poetic bent and in one of his early hold-ups he left this verse with every line written in a different handwriting style, in a rifled strongbox.

I've laboured long and hard for bread
For honour and for riches
But on my corns too long you've tread
You fine haired Sons of Bitches.

When Bart was released from prison, he told some waiting reporters he was through with crime. One of the newspapermen asked if he would keep writing poetry. "I've told you", Black Bart/Charles Bolton replied with a smile, "I'm through with crime."

HORSEBORNE ENDURANCE FEATS

*He rode all night and he steered his course
by the shining stars with a bushman's skill
and every time that he pressed his horse
"The Swagman" answered him gamely still.
He neared his home as the east was bright.
The doctor met him outside the town.
"Carew, how far did you come last night?"
"A hundred miles since the sun went down.".*

Banjo Paterson

Buffalo Bill and West Fraser's Rides

Before the age of modern communication, motor vehicles and the infrastructure to support them, many horseback journeys were made in times of emergency that seem incredible by today's standards. Other individuals have carried out monumental pilgrimages in the pursuit of land, minerals, business and mail delivery.

Perhaps the two best known long distance rides prompted by emergency in America and Australia, were both carried out by teenage boys who by circumstance, had relays of horses at their disposal.

At the age of fifteen, William Frederick Cody who was destined to be a legend of the Old West as 'Buffalo Bill' was employed by Russell Majors and Waddell who established The Pony Express Service capable of carrying mail east-west across America in about eight days [1]. Bill was given a regular run from Red Buttes on the North Platte River in Wyoming West to Three Crossings on the Sweetwater and return, a distance of 121 kilometres each way with horse changes about every thirty kilometres.

One day in 1861 Bill Cody completed his stage to Three Crossings to find the rider for the next stage had been mortally wounded in an Indian attack. There was nothing for Bill to do but carry on with the mail, and he mounted a fresh horse and set off on the 136 kilometres journey to the end of the dead rider's stage.

Bill made the journey and the return trip to Three Crossings, then without stopping undertook his own return ride to the North Platte, a total trip of 523 kilometres stopping only to throw the locked mail pouches over the saddle of each of the twenty-one horses he rode. The distance Cody completed in twenty-three hours was bettered by another Pony Express rider named 'Pony Bob' Haslam, who in similar circumstances, rode 610 kilometres in thirty-six hours. Haslam's ride is reputed to be the longest made by a Pony Express rider.

At first light on 27th October 1857 Jiman Aborigines in Queensland's Dawson River country attacked the homestead at *Hornetbank* Station killing eleven members and employees of the Fraser family and setting the scene for a ride of epic proportions by fourteen year old Sylvester 'West' Fraser.

'West' Fraser had been stunned by a blow from a club in the first moments of the raid and after the tribesmen left, he walked cautiously to *Eurombah*, a neighbouring property, to give the alarm. The story of the *Hornetbank* massacre is told in chapter 3 so we will not dwell on details here, but after the burial of the victims, 'West' Fraser took the track to Limestone Hill now known as Ipswich, on horseback despite the fact he was suffering from an ugly scalp wound.

The Dawson country was the edge of settlement in 1857, and Ipswich was the major supply point for the new squatting districts. 'West' Fraser's eldest and now only brother, William was there to pick up supplies with bullock drays. Obviously the only thought in the young lad's mind was to 'get William' and against the advice of neighbours, and offers to make the trip by others, he spurred his way south-east.

By following the bush tracks that passed as a thoroughfare from Ipswich to the extent of settlement, the distraught youth was able to exchange horses along the way at towns and stations as he carried his dreadful tidings to the only other surviving member of his family.

Three days after he left the Dawson, 'West' Fraser reeling in the saddle from exhaustion and shock, rode down the dusty street to where his brother was loading drays for the return journey. He had travelled 514 kilometres almost without stopping and after a brief respite to allow 'West' to eat and rest, and William to make arrangements for the transport of his goods, the brothers set out and made the return in almost as quick time as 'West' had come down in.

Other Australian Feats

When Fever Hits, the Doctor Rides

Roebourne in the north west of Western Australia was very much a frontier outpost in the 1890s. Dr. Frizell, a cheerful, slightly built, Irish bachelor, was the Government Medical Officer there and his jurisdiction covered a radius of some 960 kilometres from the town [2]. The only medical man in this vast area without even a qualified nurse for support, Frizell was often called on to make long, hard rides when sickness occurred in the outlying districts.

One evening a weary messenger dismounted outside Frizell's dwelling with the news that a child had died of fever at *Wilkinsons* Station and both parents were down as well. *Wilkinsons* was 224 kilometres away and relays of horses had been arranged for the doctor.

Frizell took the lightweight, skeleton, medical kit he kept prepared for such eventualities, saddled his personal mount and pushed out in the darkness heading for his first horse change. The dust of the Pilbara region plumed out behind in the moonlight as the hard riding Irishman threw the miles behind him.

From station to frontier hamlet and on again the doctor pounded by night and day stopping only to place his bag and saddle on a fresh horse and spur on towards *Wilkinsons* Station.

Frizell reached his destination fortunately in time to ensure that both parents would recover and after a meal and a few hours sleep he was in the saddle again.

Roebourne's resident engineer William Lambden Owen, Mr. J.B. Percy Manager of the Union Bank, and William Atkins railway construction contractor, were relaxing in the evening the following day when a worn out horse and rider stumbled to a staggering halt outside the doctor's residence. Dr. Frizell had been out of the saddle for only a few hours of the last twenty-nine and had ridden almost 450 kilometres.

The Doctor is reported to have made little of his feat of endurance and only remarked how thankful he was to have reached *Wilkinsons* in time to save the unfortunate child's parents.

Billy Mateer's Race Against the Flood

Dorothea McKellar's classic Australian poem '*My Country*' celebrates the wild fluctuations of weather on the island continent of Australia. Had she written her famous line about droughts and flooding rains eleven years earlier, it may well have been uppermost in the thoughts of an Eidsvold, Queensland stockman named Billy Mateer as he tightened the girths on a horse called '*Lunatic*' in cyclonic conditions on the east bank of the Brisbane River, on 18th February 1893.

1893 is remembered as the worst year for floods in the Brisbane area since white settlement and the three deluges that threatened Queensland's capital all came down in the month of February.

The first cyclone crossed the coast near Yeppoon on 1st February starting the month with more of a splash than a bang when thirty-six inches of rain fell in the headwaters of the Stanley and Brisbane Rivers in twenty-four hours. Another followed hard on its heels on 11th February and before the damage could be fully assessed let alone repaired, a third came roaring in off the Coral Sea near Bundaberg to pour its tropical rain depression on the already saturated Brisbane Valley on 17th February [3].

With creeks and rivers already running high and the country waterlogged, there was only one place for the water to go – downstream. Downstream it went and residents of the Upper Brisbane Valley saw the need to get a warning to the city of Brisbane further downstream, that the river was rising rapidly and the flood wave could get there in a few hours. The problem was the telegraph line between Esk and Ipswich was down.

Grazier Henry Plantagenet Somerset, had his *Caboonbah* Station homestead near the junction of the Brisbane and Stanley Rivers and it was he who decided that if they could get a horse and rider over the river, it would be possible to cross the D'Aguilar Range and wire news of the impending disaster to Brisbane from the telegraph office at Petrie [4].

Somerset had no trouble deciding who would be the right person to make the ride as a game stockman named Billy Mateer from *Dalgangal* Station near Eidsvold was sitting out the floods at *Caboonbah*. Mateer was a noted horseman with the all important ability to 'ride light' in the saddle, that would be so important in the gruelling test he and his mount faced.

Henry Somerset rowed his boat called '*The Daisy*' into the Brisbane River below where the flooded Stanley joined it and Bill Mateer sat in the stern handling the long lead lines attached to two of Somerset's best horses. The animals chosen were of Somerset's own breed descended from English hunters. *Oracle* by *Khedive* (imported), and *Lunatic* by *Gostwick* son of *Kelpie* (imported).

As the boat nosed into the current, Billy played out the lead ropes and quietly urged the horses to take to the water. Hesitantly with much head tossing and snorting, they followed the horseman's guidance and struck out in the wake of '*The Daisy*'.

As Henry Somerset put his strength to the oars, *Oracle* panicked and swung back to shore pulling the lead rope from Billy's hands, but *Lunatic* the second generation Australian, committed himself to the swim and surged powerfully forward after the boat. Through a maze of eddies and floating debris, the two men worked their way to the eastern bank where they landed and *Lunatic* scrambled from the river.

Billy Mateer saddled his mount in the driving rain and with oilskin buttoned to his neck and hat pulled down, he briefly clasped Henry Somerset's hand and swung into the saddle.

Bill and *Lunatic* followed Reedy Creek in their bid to reach their first goal, the crest of the D'Aguilar Range. The nature of the country determined they must swim the smaller waterway and experienced bushman that he was, Mateer would have undone the join in his reins and run up his stirrup irons to make sure nothing fouled *Lunatic's* legs if he and Billy were separated during the swim.

Lunatic faced his second crossing as gamely as the first and soon the pair were across and thrashing through the mud as they began the ascent of a steep spur that led to the top of the D'Aguilar. Billy threw his weight as far forward as he could by standing in his stirrups and leaning along the horse's neck as *Lunatic's* clawing fore feet and driving quarters propelled them upward through the rocks and trees until he stood blowing hard, on the mountain divide.

Keeping up the best pace the conditions would allow, Mateer and *Lunatic* slithered and slid down through a maze of scrubby gullies in the Mount Pleasant area until they reached the North Pine River which he followed to what was to become the site of present day Dayboro. Bill then worked his way east through the rolling hills until he pulled up outside the telegraph station at Petrie, present day North Pine.

The warning brought by *Lunatic* and Billy Mateer was in time to greatly reduce the flood damage and potential loss of life to Brisbane by the last of the 1893 floods. The sixty-five odd kilometre journey that Billy and the game horse climbed, descended, waded and swam in atrocious conditions, are a tribute to this Australian stockman and the horse that he rode so well.

Billy Mateer passed away from lung cancer in November 1934 and a plaque has been erected near Somerset Dam on the Stanley River commemorating the feat he, Henry Somerset and *Lunatic* performed in 1893.

The Extraordinary Journey of the 'Stringybark Fox'

Perhaps the most incredible, long-distance, horseborne trek anywhere was carried out by a self-effacing miner and prospector named John Dickie, known as 'The Stringybark Fox'.

The son of a Scottish farmer, John Dickie was born in the Parish of Monymusk, Aberdeenshire on 22nd August 1848 to Henry and Elizabeth Dickie (nee Fasken). Young John was baptised in September of the same year when his father took up the tenancy of *Tillyronach Farm* in the neighbouring Parish of Midmar [5].

The Dickies were a large family John having three half-brothers by his father's first marriage, and a half-sister born to Elizabeth prior to her marriage to Henry. John was the first child born to the couple and the union produced another eight – four boys and four girls. The struggle of keeping a large family on the income produced by a small tenant farm must have proved too much of a trial for Henry Dickie and sometime around 1860 he moved his family to a slum area of Aberdeen and took employment as a brewery labourer.

The move to a poor man's existence in the city proved a bad one and Henry Dickie fell ill with a fever and died in 1863 at the age of forty-five. This was to be John Dickie's first brush with fever but certainly not his last. The disease in its various forms was to have a great bearing on his life costing him loved ones and business opportunities.

John Dickie's life is largely a blank between his father's death and 1st May 1872 when he is first recorded in Australia at Mackay, Queensland working as a bailiff for Hugh McCready Junior of Palmyra Estate, Mackay. McCready, John's future brother-in-law, obtained his sugar growing expertise in the West Indies and was a leading pioneer of Queensland's Sugar Industry.

It can be assumed John received an education because he corresponded with the Queensland Land and Mine Departments, and a letter he wrote to the Editor of the *Cooktown Independent* shows he was well versed in the use of language. James Venture Mulligan, the founder of the Palmer River Goldfield was a friend of Dickie's and spoke of John as being well read. He is reported to have carried a small library of favourite books mostly of a scientific nature, by packhorse on some of his extensive wanderings.

John Dickie took up his father's profession of farming in the new country and the 500 acres he selected near Mackay in 1873 must have made him feel like landed gentry compared to his experiences growing up on a tiny tenant croft in Scotland. He worked hard at improving his land but due to a misunderstanding with authority over rent owing on the property, Dickie suffered forfeiture of his holding and lost his dream of land ownership.

North Queensland historian Jim McJannett, believes this was the first cog in the wheel that changed John Dickie from a hopeful, outgoing, young immigrant to the reclusive, lone wolf he became later.

Dickie's first known venture into mining appears to have begun then as he held miner's rights on the Palmer, Charters Towers and Hodgkinson gold fields between 1874 and 1877.

On 8th May 1875 John married Grace McCready at Mackay, Queensland. The McCreadys as told previously, were a pioneer sugar growing family in North Queensland and they considered their daughter had married beneath her. John Dickie loved his new bride and although somewhat frowned upon by his high minded in-laws, he continued working the land as a tenant farmer and for a brief period in 1875 he held the licence of the Victoria Hotel in Mackay.

On 17th April 1877 a daughter, Grace, was born to the couple at Mackay and shortly after that John moved his family to the Port Douglas area where he took work as a timber cutter and pit sawyer.

Dickie tried farming again when he paid a year's rent on a 160 acre selection at Mourilyan in the Parish of South Johnstone in 1883. Recurring bouts of the fever that would cost John so much, prevented him actively working his new block, and the Mourilyan venture failed. It was two years later when the Dickies were one of the very first residents of the remote, jungle-clad Clump Point area then part of the Cardwell district, that catastrophe overtook the thirty-seven year old Scotsman and his young daughter.

On 8th January 1885 John's beloved wife Grace died of fever, that insidious sickness that laid so many of Australia's pioneers low in the tropical north. John Dickie acted first as nurse and doctor then dug a grave and read a eulogy over the girl who had won his heart while their eight year old daughter watched helplessly.

This was the turning point for Dickie and in the depths of despair he turned from a convivial, young member of an emerging rural society to the withdrawn isolationist who would become known as 'The Stringybark Fox'.

John Dickie was in the depths of black depression when he arrived in Cardwell to report his Grace's death. He gave his daughter's name to Thomas Young who took the particulars, not as Grace as it should have been, but by the pet name 'Betsy' that he always called her. The nickname 'Betsy' came from the child's two grandmothers Elizabeth Dickie nee Fasken, and Elizabeth McCready nee Doak, and was apparently the title Grace was known by in the family. We can only guess at John's state of mind at the time as he stumbled through some sort of arrangements for his daughter's welfare before fleeing to the bush and the solitude.

From that time on only sketchy details are available about the life of Grace 'Betsy' Dickie. It could not have been easy for a young girl cast adrift from her family in those times. It was not practical for John to raise the child himself and the attitudes of the time would have been against a single father.

Her proud relations on her mother's side obviously wanted nothing to do with the offspring of the daughter they rejected, and we must assume Grace lived in foster homes until adulthood. She married on 6th September 1897 in Brisbane. Her husband Lyndon J.A. Poingdestre late of the Queensland Native Mounted Police was ten years older than her father, which suggests her uncertain early years made her long for a father figure. Her husband died in 1924 and in 1928 she married again to a man named Hiscock. Grace Dickie (Poingdestre/Hiscock) passed away childless in 1956 in Roma, Queensland.

By 1885 John Dickie had lost interest in his selection and was tin prospecting and/or mining near Herberton on the Atherton Tablelands. John Dickie's heritage was the land and he loved it. It seems sure his ambition was farm and family and his previous ventures into mining were directed towards raising the money to improve his selection. Following the tragic breakdown of his dreams, prospecting and mining as a profession, claimed the rest of his working life.

He is recorded as holding gold claims and/or leases at Croydon in the Gulf Country between 1886 and 1887 and in late 1887 he began his first long distance prospecting trip in the wild lands of Cape York Peninsula.

Captain William Campbell Thomson of the Australian Steam Navigation Company dropped Dickie off at Newcastle Bay just below the eastern tip of Cape York. He landed with a piece of calico, a blanket, gun, quartpot, tea, sugar and a bag of oatmeal. Other sources name his drop off point as Cape Weymouth, but concrete evidence exists that the '*City of Melbourne*' put him ashore at Newcastle Bay.

He worked his way south on foot and during the trip found traces of gold, a wolfram find that a field would result from, as well as tin, coal, fluorite and molybdenite. He worked his way down the Cape eventually arriving ragged and barefoot at Coen some 350 kilometres south as the crow flies, but Dickie would have travelled a lot further considering the perambulations of a prospector. Jim McJannett who knows more about John Dickie than any other person, has tracked his Cape York journey from Newcastle Bay to the Pascoe River on to Fair Cape and then by boat to Cape Grenville. From Cape Grenville he walked to Coen then north again to Temple Bay. He again hitched a ride by boat to his starting point at Newcastle Bay then once again he trekked south to Coen and on to Cooktown via Laura. A long, long walk by anyone's standards, 1,359 kilometres in beeline calculations, but all in the day's work for the 'Stringybark Fox' and a stroll in the park compared to the horseback pilgrimage he would undertake in a few years time.

Until 1894 other than a short period prospecting a little south of Cairns, Dickie prospected throughout The Cape York area with some success but it was during that year the odyssey that was to prove him as one of Australia's finest bushmen, began. In March of 1894 John Dickie saddled his favourite mount 'Tam'o'Shanter', loaded two other packhorses which had come from the Cape with him, and left Croydon heading for the Old Gulf Track first traversed by the explorer Ludwig Leichhardt in 1845, and opened as a stock-route by Wentworth D'Arcy Uhr in 1872.

By July he was recorded over the border of Western Australia when Constable John Caddon recorded the brands of his horses at the Twelve Mile Camp between Flora Valley and Halls Creek. John worked his way further down through the west and was in the area near the Murchison River that would become known as Meekatharra, when he came across the bleached skull of an aborigine in the fork of a tree that appeared to glow in the light of the moon.

An idea of how to greatly reduce the risk of travel in regions populated by wild aborigines began forming in his mind, an idea prompted by glowing fungi seen in the Cape York jungles, and when he reached the Kalgoorlie/Coolgardie Goldfields he consulted a chemist named John Boileau. 'Long John' Boileau was a Dublin Irishman who arrived at the West Australian Goldfields seeking his fortune. He found a ready market for his profession as a chemist and when John Dickie explained what he had in mind, Boileau set to work. He came up with an oil-based paste that shone in the dark with an eerie glow, perfect for Dickie's purpose.

Boileau's potion was used to smear around the teeth and eye sockets of the now blackened skull Dickie had found in the Murchison River area. Suspended on a fishing line and swivel at night, the skull would turn slowly causing the phosphorescent death's head to appear and disappear in the darkness.

Not only did it terrify the wild Aborigines who came to fear John Dickie more than their own tribal shamans or witchdoctors, leaving him and his camps strictly alone, it almost scared the life out of some of his white visitors as well. When Long John Boileau's paste ran out, Dickie improvised his own concoction with fungi, goanna oil and other secret ingredients. John Dickie's 'brew' became part of the folklore of mining districts and allowed him to traverse regions from which other whites had never returned.

Prior to the development of his brew, John had relied on acting like a crazy man to avoid confrontation with natives. Aborigines generally respected mental illness and would do nothing to harm anyone they saw as mad, due to their belief they would inherit the traits of those they killed and in some cases, ate. There is little doubt that John did not regard playing the 'loon' as a foolproof defence however. You cannot 'bung on an act' while sleeping, and he obviously placed more faith in his glowing paste than his acting ability.

Some time during his sojourn in the west, John decided to pack up his three faithful horses and make a trip west to the coast at Fremantle to catch up with an old mate. He arrived to find his friend with a contract to build a large shed and suffering from a broken arm. Dickie picked up the woodworking tools in a spirit of mateship and helped finish the job.

Jim McJannett believes he contacted Grace his daughter while there, as on her marriage certificate dated 6th September 1897, she listed her father's occupation as a carpenter rather than a miner or prospector.

By late 1895 or early 1896, John had undertaken the 600 odd kilometre trip back to the Goldfields and located alluvial gold near a find made by Bill McJannett at Darlot near Coolgardie. Miners rushed to the area from New Zealand as well as all parts of Australia and the gold Dickie won there allowed him to finance the most ambitious part of his trans-Australian ramblings yet undertaken.

In the winter of 1896, the 'Stringybark Fox' packed up his horses, left Darlot and wended his way north-east into the dead heart of the continent. Attempts at travel through desperate regions between the Great Victoria and Gibson Deserts had turned back competent travellers using camels, when attempted from either east or west, and it is incredible to think that John Dickie succeeded with horses where the ships of the desert had failed.

The route he took is not known for sure, but the most logical way would have been via the Warburton region. Dickie packed all three of his horses and walked until he had lightened his food supplies enough to allow him to ride, but even then, ever mindful of sparing his animals, he spread the load and walked for long distances. There is no record of exactly what he took, but we can assume there were no luxuries like books on that trip, and one horse was sure to have carried water canteens.

John Dickie told the author Ion Idriess that whenever he found water in the desert regions, he would soak himself full sip by sip until he could hold no more encouraging his horses to do the same. He would then travel at night and lie up in the heat of the day to conserve his horses and his own energy.

The bushcraft needed to locate jealously guarded native wells, was phenomenal. Dickie would doubtlessly have followed aborigine's tracks and smoke signals, and the presence of small birds and insects that could not travel far from water, would all have contributed to him winning the Russian Roulette game he played with the desert. One mistake was and still is, all it takes to fail in those almost waterless regions, and to fail is to perish.

As he worked his way towards a gap in a range west of Alice Springs one day, John became aware of armed Aboriginal warriors in large numbers keeping pace with him from the high ground on either side. Night had fallen before he reached his objective, so the wily prospector set up his camp and included a few surprises for the watchers from the heights.

Some folding Chinese paper balls were hung in trees together with a camel's skull both suitably decorated with 'Dickie's Brew'. The skull was attached to a rope thrown over a tree limb. John hauled the glowing, grinning apparition suddenly into the air and exploded some half plugs of dynamite at the same time. This was more than enough to convince the would-be attackers that a malevolent spirit with a loud and angry voice was sharing the camp of their intended victim.

John Dickie slept soundly that night and when scouting around next morning, he found spears, dilly-bags and provisions that had been hurriedly abandoned as the natives bolted helter-skelter to escape the 'debil-debil'.

Dickie continued on his way and eventually he and his three remarkable horses walked out of the desert at Alice Springs, which was then little more than a rustic settlement and a telegraph station.

John Dickie had achieved the impossible. He and his three equine companions had made a trek over at least 1,800 kilometres of the most inhospitable country on earth. By a superlative feat of bush and horsemanship, 'The Stringybark Fox' had done what no horseborne traveller has achieved before or since but the journey was far from over.

John Dickie carried on from Alice Springs across the Territory border to the Queensland copper mining and cattle town of Cloncurry. He mined there for a short time after finding a small pocket of gold about ninety kilometres from 'The Curry'. With enough money from his find and the sale of his claim, to restock his packs and buy a few other essentials, John Dickie and his three exceptional horses, struck out to the east.

The vast grassland plains at the head of the Gulf Country rivers fell away beneath his horses' feet and after reaching the Dividing Range, John passed through the North Queensland mining towns of Charters Towers and Ravenswood before reaching the coast near Mackay.

John Dickie and his horses had crossed Australia east to west, and west to east through some of the most savage country on the face of the earth. He and his faithful animals had largely lived on what the country provided with John eating small game and even snakes and lizards. If all edible vegetation failed, he would have tightened his belt a notch and fed some of the precious oatmeal he swore by as a traveller's staple, to his horses so they could make the long traverses between waters, many kilometres apart.

John worked for a while at Edmonton south of Cairns supervising the laying of sugar cane rail tracks on his way back to Cape York. By 1898 he was prospecting on the Morehead and Coleman Rivers under a Government sponsorship. Somewhere in the Coleman River district John Dickie discovered a very rich gold reef or reefs, but when he tried to return some time later, a cyclone had been through the area. With many square kilometres of trees flattened, the country was practically impassable. Long grass had grown up and Dickie could not relocate his find.

In late 1899 John Dickie discovered the Hamilton Goldfield, which he reported at Coen on 2nd January 1900. Following the decline of the Palmer and Hodgkinson fields, Cooktown was fading as a supply centre for the mines and Dickie's alluvial find breathed new life into the town. What a wonderful gift to begin a new century.

Not content with just an alluvial field, he poked around the ridges and located the mother lode as well and on 27th January, less than a month later, John Dickie reported the Hamilton Reefing Field. By the end of March, the 'Stringybark Fox' was gone. His work as a prospector was done. The Hamilton Goldfield and Ebagoolah, the town that sprung from it, held no further attraction for John Dickie and once again, he took to the bush in his lonely quest for new finds.

Until 1919 John Dickie poked about his beloved Cape York prospecting. In 1910 he led an expedition in search of gold on the northern part of Cape York that was funded in part by the Government. His companions were James Dick and Arthur H.Sheffield on the almost six month trip and it is due to the recollections of Sheffield who at the age of ninety-one was interviewed by Jim McJannett, that much of the 'Stringybark Fox's' history has remained.

That Dickie was a loner and an introvert is beyond question, but he was no anti-social 'hatter' like some of the misfits who roamed the outback. If he rode up to someone's camp, no one enjoyed a yarn more than 'The Stringybark Fox'. Well read, he could quote Shakespeare and the Scriptures, was a great follower of Robbie Burns as well as being quite a fan of a relatively new Australian poet named Henry Lawson.

Those who knew him were in awe of his ability as a prospector and it was a common thing for Peninsula residents to declare that "Old John Dickie can smell out gold." Dickie's prospecting and bush survival skills were all the more exceptional when we consider that he was not born to the Australian bush and had made his first long trips in arid country when approaching middle age.

However, this man who could and would speak freely and fluently on almost any subject raised around a Cape York campfire, was like a cautious clam if questioned about his own life and achievements. It is only through the efforts of others like Ion Idriess and Jim McJannett that John Dickie's story has survived. Sheffield told McJannett that it was as if he had no past prior to his arrival on Cape York.

He made several other significant mineral finds including the Alice River or Philp Goldfield, and an antimony find he named The Coghlan, but he could never re-locate the rich reef he found in the Coleman River country in 1898. It was because of this, that some people who probably would have found it difficult to follow John Dickie down the main street of one of the 'rushers'' camps and boom towns that sprung up in his wake, labelled him a poor bushman. Somehow this unfair tag stuck to the 'Stringybark Fox' in later life but it was nonsense as his achievements spoke for themselves.

People who knew John said his sense of direction was not perfect but John Dickie was proof of the old saying 'if you don't care where you are, you're not lost'. Almost all good bushmen get 'slewed' now and then and Dickie would always find his way eventually.

A pathfinder is only as good as the animals he rides and packs, and there is no doubt John Dickie loved and cared for the horses that served him so well. His favourite mount 'Tam'o'Shanter' which crossed so much of the Australian continent with him, reached a big age for a horse and passed away in retirement at Mount Cook, Cooktown shortly after World War I.

John arrived on the Annan River tin fields near Cooktown about 1913. He had been on yet another prospecting jaunt up the Cape using packhorses supplied by the government. One had died. In much the same way as the South Australian Government made that other intrepid pathfinder, Nat Buchanan, pay £40 for a camel loaned to him, which died on his Tanami Desert Expedition in 1896, the powers that be in Queensland were after their pound of horseflesh from John Dickie. The sixty five year old took a job on wages with the Annan River Tin Mining Company to pay the debt, but his old Peninsula mates sent 'round the hat and squared him with his ungrateful government, a government that only paid the prospector £500 as his reward for finding the Hamilton Goldfield when it should have been £1,000.

One cannot help wondering if the penny-pinching bureaucrats who pursued Dickie over the dead packhorse, reimbursed the State for their broken pencils and wasted paper. It is much more likely they diverted some of the duties raised

John Dickie 'The Stringybark Fox' looks every inch the frontier prospector in this 1912 photo. The heavy holstered revolver on Dickie's right hip was for defence if needed, and the smaller long barrelled firearm under his left wrist, was John's meat gun. He was reported to be a deadly shot and the way the guns are carried would indicate John Dickie fired with his left hand.

Photo courtesy J. McJannett believed taken by Summerhayes Studio, Cooktown)

via John's mineral finds to cover their losses.

Dickie stayed at Annan River long enough to fill his tucker-bag and put a few quid in his kick (money in his pocket) and headed west again. After all this time on Cape York, he prospected out as far as Cloncurry on this trip via Croydon and Richmond.

On 17th April 1918 John and a younger man named James Hare, left Richmond for another try at locating Dickie's elusive Coleman River Reef. The pair became separated in a search for water but they were bushmen, and eventually they both made their way to habitation knocked up, dry and hungry, but alive. Hare made it to *Gledswood* Station in the head of the Norman River and John to a Chinese dwelling on the Gilbert River.

After this ordeal, the old man headed south to Innisfail in 1919 to take things easy and have a bit of a spell, but by 1920 the indomitable Dickie was back on the Cape looking for his lost reef. He spent the years from 1920 to 1923 in a fruitless search for the rich deposit he had located and lost again but by 1924 his health was failing and he again travelled south to Innisfail.

John Dickie's enthusiasm was unflagging but the years had finally caught up with the tireless prospector. Just a week before departing on yet another search for his Coleman reef, he suddenly fell ill with heart disease and within the day, John Dickie was dead. He passed away on 31st October 1924 in the Innisfail Hospital aged seventy-six years, a big age in those days for one who had endured so very much hardship.

It seems rather ironic that a man who averted death or injury so many times from hostile natives through the use of hocus-pocus to arouse their fear of the supernatural, should go to his Maker on Halloween. He was buried in the Innisfail cemetery the following day in a pauper's grave. Apparently no obituary was printed following Dickie's death, but that would have been of small concern to the 'Stringybark Fox'. He'd already had

JOHN DICKIE
1848 – 1924 76 YEARS
A SCOTSMAN KNOWN AS
"THE STRINGYBARK FOX"
DISCOVERER OF EBAGOOLAH GOLD FIELD
CAPE YORK PENINSULA

Almost shaded by a line of trees, John Dickie's final place of rest shows prominent as the only marked grave in the public (paupers) section of the Innisfail Cemetery. Little but birdsong and the movement of leaves enters his sleep, a peacefulness, a loneliness befitting a lone-wolf prospector, the man whom all on Cape York Peninsula knew as 'The Stringybark Fox'.

(Photo and caption courtesy Jim McJannett.)

several printed while alive! When overdue on his ramblings, he was at least three times, reported missing presumed dead.

John Dickie, the 'Stringybark Fox' should have gone down in Australia's history along-side men like Nat Buchanan as an explorer, pathfinder and opener of new country. His bushmanship and ability in the wild lands were second to none and it is a shame he has never received the recognition he deserves in the annals of Australia.

Footnote: I would not have been able to recount the story of John Dickie were it not for the years of study of this remarkable Australian, by Cairns historian, Jim McJannett who generously allowed me access to his material. In over forty years of research, Jim has pieced the story of the 'Stringybark Fox' together ensuring that a wonderful piece of history is preserved. It is largely due to Jim that a plaque has now been placed on Dickie's grave to commemorate a tiny part of his achievements. Jim is still actively researching the life of John Dickie and welcomes any input. He can be contacted by email at ebagoola@hotmail.com

American Feats of Endurance

How Louis Remme Outrode Bankruptcy

In late 1854, Cattle Rancher, Louis Remme was in Sacramento, California feeling pretty pleased with himself. The day before, he had sold a herd for the pricely sum of $12,500 – a lot of money in the mid 19th Century [6]. Remme deposited his recently acquired wealth in stagecoach company Adams and Co's, Express Office, and sat back for a few days of well-earned rest and recreation.

Reclining in a comfortable chair in the lobby of his hotel, casually leafing through the latest issue of *The Sacremento Union*, Lois Remme came across a news item that brought him upright with a jerk. "Adams and Company close doors. Managing director Woods has flown."

All thoughts of holidaying forgotten, Remme hit the floor running. On reaching the office to present his receipt his heart sank as he saw the angry mob already there, demanding their savings. What could he do? The news had come via San Francisco, so there would be no chance of drawing his money there. He needed to get to a branch that was as yet unaware of the crash and he needed to do it fast.

Remme's agile brain was working overtime. Portland, Oregon was his best bet. The news would get there with the steamer due to leave San Francisco the following day. It was worth a try.

He headed straight for the Sacremento River where he boarded a paddle steamer for Knights Landing sixty-seven kilometres up river. At Knights Landing Louis borrowed a horse which he rode to his friend Judge Diefendorf's place at the head of Grand Island, and after getting a loan of a good sort of mount from the Judge, Remme headed north at the gallop.

Mile by mile Louis Remme drove northwards. It was fortunate for him he was well known in the north west and he had little trouble borrowing good quality horseflesh. He rode hard and changed mounts as often as he could. In seventy hours he was at Yreka not far below the Oregon border and hoping against hope, his rival the steamer, was making heavy work of its coastal passage.

Five days had passed and he was well into Oregon when he reached the town of Eugene. The following morning he pounded into French Prairie so tired he could not stay awake. Remme mumbled his request for another horse and his friend in French Prairie seeing his exhausted condition provided him with a horse recently purchased in Portland that would head that way with no guidance from its rider.

Louis had snatched a mere ten hours sleep in the 140 odd he had ridden, and in the final leg of his trip he managed to keep his seat despite dozing off constantly. It was mid morning when the exhausted cattleman rode into Portland.

He doused his head in a horse trough and after attempting to brush himself down and achieve at least a trace of decorum Remme strode through the door of Adams and Company and laid his receipt on the counter.

The clerk checked the receipt and with agonizing slowness, to Louis at least, began counting out the gold coins. The morning sun was high when Louis Remme walked out of Adams and Company's Portland office with almost nineteen kilograms of pure gold in a gunny-sack.

The relieved rancher had only got a small way down the street when he heard the boom of a cannon signalling the arrival of the '*Columbia*' up the river of the same name, with the news of Adams' failure.

Remme saw to the care of his horse and went looking for a place to sleep. In 143 hours he had ridden 1,070 kilometres with only ten hours sleep but financial crash or not, the gutsy cattleman was still in business.

From the Arctic to the Tropics

In the last year of the 19th Century an expatriate Englishman named Roger Ashwell Pocock set out on an amazing journey. Pocock was the son of a naval man who had migrated to Canada in 1882. He worked as a survey hand, prospector, seaman, cowboy and as a civilian scout for the Royal Canadian Mounted Police [7].

Some time in 1899 Roger Pocock set out on horseback from Fort McLeod in Alberta to ride to Mexico City. He completed the 5,800 kilometres in 147 days travelling through regions in the western United States where the menace of white outlaws and the disgruntled remnants of the plains tribes were still very real.

Following his epic ride, Pocock sailed to England where he enlisted for the Boer War. He fought in the South African conflict attaining the rank of Corporal.

The Sheridan to Galena Ride

In 1897 in America, Dr. William A Bruett, special Commissioner of the Bureau of Animal Industry put the wheels in motion for the running of a 3,860 kilometre ride between Sheridan, Wyoming and Galena, Illinois with the idea of promoting western horses for export as army remounts [8].

Two cowboy brothers named Bill and Bert Gabriel caught two Wyoming range horses with a good dash of mustang blood line, broke them to saddle and on 5th June they started for Galena.

The horses were fed no grain and made do with grass along the route. They also did the entire trip without shoes on their rock-hard, range hooves. The two rugged cowboys camped with their horses at night and 91 days later on 6th September they rode their now well-broken broncos into Galena.

When the feat is analysed it was not really an exceptional achievement. The average distance travelled each day by the Gabriels and their horses, was about forty-two kilometres . This would allow plenty of time for rest and grazing and the daily distance could have been accomplished in seven to eight hours without travelling faster than a walk. As Americans often travelled at a slow trot called a running walk by them and known as a jig jog in Australia, they may well have shortened their daily times more with little impact on the horses.

Knowledgeable horsemen faced with making one horse last a long distance but still concerned with time, will travel at a trot. Trotting is a two beat gait where impact with the ground is reduced by two feet striking together. It is harder on the rider, but easy on the horse and in most cases where records of distance and endurance have been made with only one mount at the rider's disposal, the trot is the preferred gait.

We can assume Bill and Bert Gabriel set a daily routine regarding distance and rest stops and simply kept poking away at the job in hand without putting undue pressure on their mounts. The country along the route was reasonably populated and it was summer/autumn so the cowboys would have got by with a blanket each and would only have had to carry sustenance type food for a day or two, and a few simple utensils.

Dr. Bruett's strategy paid off however, with a marked increase in export sales following the publicity generated by Bill and Bert's ride. Previously the average number of horses exported per year was around the 500 mark. Following the Sheridan-Galena ride and the subsequent ballyhoo by its promoters, 15,000 western bred horses were knocked down to foreign army buyers at the Chicago sales shortly afterwards in late1897.

Australia was heavily involved in the remount trade during this period as well, and the Australian bred Walers were regarded as second to none by Army buyers. For endurance and weight carrying plus the ability to hold condition in hard country, the Australian Waler had no equals and was the most sought after warhorse in the world.

BUCKJUMPERS & ROUGHRIDERS

*The blokes who claim they rode that 'orse,
if yer got 'em mustered up,
would outnumber the attendance
at the bloody Melbourne cup.
They never cease to tell yer
'bout their deeds of long ago.
They think they've got yer kidded,
but the knowin' blokes all know.
And still around those western towns
where mornin' suns rise red,
yer'll find the windy skite who claims
he once rode 'Rocky Ned'.*

Keith Garvey

The Reason for the Legend

Anyone who has ever reined up a sullen, lop-eared colt with a jammed tail and a kink in its backbone, in the half light of dawn, stuck a boot in the stirrup, a knee behind the shoulder and swung up knowing full well it was 'on' for young and old, will understand why the man on the bucking horse is one of cattle country's most enduring images.

Nothing gave a man from the back-country more respect from his peers than the ability to ride a rough one. When some wiry individual in riding boots propping up a bar in a cattle town or squatting on his heels by a campfire offered the opinion that 'so and so could ride', the translation to those who knew, was that the subject of the discussion could win any rough riding contest.

With the same laconic understatement of people who live and work close to the land world wide, the observation that 'such and such a horse could buck' meant that only the select few who could literally ride anything with hair on, would have a chance of staying on top of that particular animal.

An old adage that has become part of horseman's lore anywhere in the world stock is handled from horseback, runs along the lines of *"There's not a horse that can't be rode and there's not a man who can't be throw'd".*

From world famous Rodeos like Calgary, Mt. Isa, Cheyenne or Warwick to the stock camps of the outback and cow camps of the western States of America, from the Asian Steppes with their hard riding Mongol horsemen to the Gauchos of the South American Pampas, the champion wild horse rider is the stuff of legends.

Many old hands will tell you what wonderful horsemen the people of their generation were, deride their counterparts of today for being unable to stay on a gate in a stiff breeze, and in the next breath come out with that old horseman's classic "*Of course there were a lot more bad 'orses 'round in them days.*"

One is tempted to ask the question, if the horsemen were so great, how come there were so many bad horses? But that line of questioning usually provokes the standard "Listen here Sonny" reaction which tends to leave both parties aggravated and nothing settled.

The reality of the matter is economics decided how horses would be handled. To stand up to the rigours of life in a stock camp a horse needed to be at least four years old. Asking any younger animal to carry a rider at the pace and for the time necessary to get stock mustered and in hand, would only result in a knocked up horse and a rider on foot.

Property owners found it easier and cheaper to do no more than wean and brand their young horses, geld the males and employ a breaker at a contract price per head to get them rideable when they were mature enough to stand the work. To make a living, a horsebreaker needed to turn out at least six horses a week. A station or ranch breaker was not expected to educate the horse to any high degree. Their job was to catch it, get it leading, mouth and ride the youngster and handle its legs well enough to enable it to be shod if working in stony country. The finer points of education were taught by the ringers or cowboys themselves as they worked stock. The highest accolade paid to anyone who worked among cattle, was to be known as a good hand with stock, who could get a young horse going.

There were always those naturally gifted handlers who could communicate their wishes to the animal with a minimum of aggravation and smooth the breaking-in process. The rest just did it the way it worked for them.

While the bosses thought nothing of handing big, wild mature horses to the breaker who was expected to have them ready for work in one week, they would have plenty to say if too much hide was knocked off them. Any horse tamer who wanted to keep working needed to be able to get the job done in the required time without physically hurting the animal.

There is a large school of thought these days that old time horse breakers were 'callous brutes' who broke the animal's spirit. Nothing could be further from the truth. Given the circumstances these men and sometimes women, worked under, they were truly gifted people to turn out the quality of stock horses they did. Admittedly the word 'horsebreaker' does conjure up visions of cruelty to the uninitiated, but it is just a term that evolved meaning a person who broke in an animal to accept a rider and/or work in harness. No broken spirited plug ever caught the lead of a mob of wild bush cattle in heavy Brigalow Scrub, or performed the exacting task of cutting out required beasts and separating them from the main herd on an open camp without benefit of stockyards.

If horses were going to be ready to turn over to the stock camp in seven days, the breaker needed to be riding them by the third or fourth day. No decent breaker encouraged bucking, but they were fairly philosophical about it if a youngster dropped its head. You didn't take the job on if you couldn't stay on.

A friend of mine who publishes a horse magazine these days after a lifetime of training and competing summed up the experienced horseman's fatalistic attitude after watching an overseas demonstrator play around for hours with a youngster that showed signs of being a bit 'tight'. "Hell Mate," he called to the trainer or clinician as he liked to be known. "Get on him! All he'll do is buck!"

Of course there were failures. When mature animals begin their working life with a seven day crash course, how could there not be? It was these failures that became the 'bad' horses old blokes love to talk about. To compare the painstaking and time consuming style of modern equine education where horses are handled from a very young age, superior though it is for horse and rider in every way, with the situation old-time breakers faced, simply shows a lack of understanding of conditions in those times.

Right or wrong, that was the way it was. The riders who took horses off the breakers could all handle touchy colts as horses undergoing initial training were known in Australia and America regardless of age or sex, and could break-in themselves if necessary. They took pride in their ability to ride a rough one and most young ringers or cowboys would have felt a bit cheated if a few of the horses assigned to them didn't buck when they were fresh.

The bosses compounded the situation even further in Australia, by keeping gentler and more talented mares at work instead of breeding from them, and using the buckjumping ratbags of uncertain temperament as brood mares. America appears to have held to a general policy of using only geldings as work horses on the big ranches.

Before the development of the American quarterhorse and the Australian stockhorse, thoroughbred sires were popular on stations and ranches and once again, economics reared its head when stallions of uncertain temperament rejected by the racing industry, often finished up on properties because they were cheap.

Many places were breeding bucking horses but no matter how tough they were to handle, there were riders equal to the task. It was from these confrontations between man and beast, that the legends of buckjump and bronc riders, and the wild steeds they straddled, were born.

Australia's Best

Lance Skuthorpe

The title of Australia's greatest roughrider has always sat squarely on the shoulders of Lance Skuthorpe senior [1]. His son Lance junior was a phenomenal rider who won worldwide accolades, at one time riding a bucking horse with a big reputation in the United States sitting in a flat Australian saddle with his arms folded, but he never eclipsed his father's achievements.

Lance senior, born Lancelot Albert at Kurrajong, New South Wales in 1870, spent his boyhood years at Garah north of Moree in New South Wales and developed a marvellous affinity with horses from the time he was lifted to the back of his first pony *Grey Billie*. Skuthorpe drifted through the Australian bush as a young man giving horse taming or gentling exhibitions.

In one of his demonstrations in the Mt. Gambier region shortly before he repeated the feat of poet Adam Lindsay Gordon, by jumping his horse over a fence on the face of a cliff and turning the animal in midair to land within two metres between the cliff and the fence, Lance performed another amazing feat.

Skuthorpe was presented with a horse that was a known striker who used his front feet to warn off anyone who invaded his personal space. Minutes into his demonstration, he picked up the animal's front leg and stroked the horse's hoof down his own forehead.

Skuthorpe published a pamphlet advocating gentleness and understanding as the key to horse training. His observations of the horse's instinct as a prey animal that would allow nothing on it's back, are being reiterated by animal behaviourists of today.

It would probably surprise many of the current crop of horse whisperers to know that Lance Skuthorpe was giving similar exhibitions to their own, over a hundred years ago.

However, Lance was a man before his time. In those days everyday people understood horses as well as their counterparts understand cars today, and gun buckjump riders, like today's racetrack and drag-strip heroes, enjoyed huge status. Lance soon found that by demonstrating the ease and gentleness of his methods, he attracted mediocre crowds who were constantly accusing him of trickery and sniffing at his gear for traces of chloroform or other calming substances. Let's face it, there'd be no point in advocating defensive driving at Bathurst or the Indy 500. Ever the Showman, Skuthorpe gave his audience what they wanted. He took on every bucker that came his way and rode with a loose limbed style that seemed to adapt effortlessly to every twist and contortion of the horse. No rider in Australia, before or since, has achieved the fame Lance did from the saddle of a bucking horse.

Lance put a buckjump show together with other supporting acts, and toured the country. When a wire walker named Martin Breheny nicknamed 'Martini' left his employ to set up his own show, it set the scene for Skuthorpe's greatest ride. Martini first acquired the notorious *Dargin's Grey* and when the nondescript horse scarcely larger than pony size, unloaded its challengers night after night, Martini's show soon began to rival Lance's 'Wild Australia' production. Martini who is reputed to have walked a tightrope across Niagara Falls, was a smart operator who had learned the secrets of running a buckjump show while touring with Harmston's American and Continental Cirque on an 1890-91 tour of Australia [2].

As the grey horse aged, Martini needed another star attraction, which he found in *Bobs*. The big, slab-shouldered horse threw everyone who tried him with almost contemptuous ease prompting Martini to make the claim that not even Lance Skuthorpe could ride him.

Lance was touring Queensland in 1906 when a racehorse owner named Cassells, offered to back him to ride *Bobs* for £1,000. Never one to back away from a challenge, he paddocked his horses near Kingaroy and travelled south to Sydney for the match. In a below street level venue at Rawson's place in Sydney on St. Patrick's Day 1906, Lance rode the horse that had thrown over eight hundred challengers.

Bobs was a master of the king-buck who at the top of his leap, would unleash a buck within a buck. Lance however, was equal to the challenge and when Australia's greatest buckjumper of the time, subsided into mild pigroots in the flare of the gaslights, over 3,000 people watched Lance Skuthorpe's triumph.

Lance lived a long and adventurous life. His show became the most famous in Australia and his fortunes went from cashed up to flat broke several times. He was a talented story-teller, reciter, author and cartoonist as well as an amazing horseman. Lance Skuthorpe died of cancer on 9th February 1958 at Bankstown in Sydney.

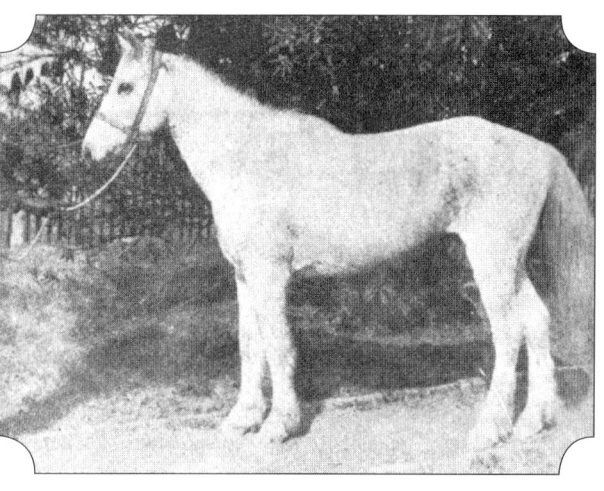

Australia's first nationally known bucking horse, *Dargin's Grey* was a rebel all his life. His story is told in this chapter.

Photo first published in *North Queensland Register*, 1905.

America's Best

Booger Red

The American Cowboy who seems to be generally recognised as the greatest bronc rider of earlier times, was a colourful character christened Samual Thomas Privett but known far and wide as 'Booger Red' [3].

Red was born on his parents' ranch in Erath County, Texas on 29th December 1864 and had made a name for himself locally as a roughrider by age twelve.

When Red was thirteen he was experimenting with some homemade fireworks when the whole box and dice blew up leaving his face badly scarred. This didn't appear to hinder the irrepressible youngster much and throughout his life he bragged he was the "ugliest man in the world".

The red headed, bronc rider's parents died when he was fifteen and Booger Red became a travelling bronco buster and rodeo performer. His riding must have been very similar to Lance Skuthorpe's with a balance rather than grip style. He was a casual individual who was in the habit of looking over his shoulder and making wisecracks to the audience from the saddle of notorious outlaws.

Where many riders favoured custom made and often intricately ornamented saddles, Red's bucking rig was pretty basic. He rode in an uncovered saddletree with girths, stirrups and skirts (padding for the horse's back) attached. Red must have reckoned that since bucking saddles get subjected to some pretty rough treatment, at least there wasn't a lot to wreck on his.

In 1895 Booger Red Privett married Molly Webb at Bronte, Texas. The couple had six children who all followed their father's footsteps becoming Wild West Show and Rodeo performers.

For many years Booger Red toured the Lone Star State with his own Bronc Show as travelling buckjump shows were known in the United States. His was always famous for having some of the rankest bucking horses in the business. To have held a job with Booger Red Privett was enough to seal the credentials of any bronc rider. Even in his fifties, Booger would not think twice about stepping on a horse that had bucked off a local challenger and fitting a ride on it.

Booger Red's last performance was a special exhibition ride at Fort Worth, Texas in March 1925. He was sixty years of age, which gave him one of the longest roughriding careers anywhere.

Samuel Thomas 'Booger Red' Privett died of natural causes at Miami, Oklahoma two weeks after his final ride.

Legendary Buckers

There have been countless great bucking horses and just as many great rides on horses that bucked like furies, but did not keep bucking hard like the legendary ones. Australia saw Tex Morton's *Aristocrat* and *Mandrake*, *Spinifex* and more recently *Blondie* of the Mt. Isa Rodeo Club, Warwick's *Arrawidgee* and *Knickerbockerbuckaroo*, Lance Skuthorpe's *Snips* and *Queenslander* as well as latter day stars like *King's Cross* campaigned by Stock Contractor Gary McPhee and Gill Brothers' *Flying Devil*.

The United States saw buckjumpers or Broncs as they called them, like *Burgett* owned by William Woods of Blackland, Texas who cost countless wagered dollars to cowboys who fancied their chances with the iron grey stallion. *Burgett* threw all comers until Jim Woods of Forney, Texas rode him to a stand in 1896.

Other notable early times American buckers like *Whistling Annie*, *Long Tom*, *Bittercreek*, *Casey Jones* and *Old Colonial* made their mark on the newly emerging Wild West Show and Rodeo Circuit.

It was the northern states like Montana and Wyoming, and Canada that produced the most awesome, American, bucking horses. Draughthorse and thoroughbred blood was introduced to range bred mustangs to produce tall horses stout enough to crash through the winter snowdrifts and the big tough range-bred buckers from the northern part of North America gained a fearsome reputation.

The Bruising of Buster Ivory

During the 1930s and 1940s tractor manufacturers in North America were offering trade-ins on horse teams as an inducement to farmers to turn mechanical. An executive from a machinery company in Canada contacted rodeo stock contractor Harry Rowell of Heywood, California with a view to selling some of the ex-team horses that were rapidly accumulating [4].

Harry sent Buster Ivory, a top rider of the time and a few of his bronc riding buddies north with instructions to try the horses out and buy the ones that bucked. Buster rang Harry a few days later from the Canadian Prairies to tell him that the plough horses had triumphed over the rodeo riders. "Hell, they all bucked! We only got a handful of 'em rode" the voice came down the phone.

The draught stock in that part of the country had been bred up from range horses crossed with English Shires then bred back to thoroughbred stallions. The toughness of the Mustang and range bred cowhorses coupled with the size and power of the shire and the spirit of the thoroughbred formed a deadly combination. In his reminiscences, Buster recalled the horses were taller than the chutes he and his mates tried to ride them out of.

Maybe those horses had got wind of the fact that the only other market for them was pet food. Perhaps they'd been happy ploughing and had no intention of adopting a career change. Whatever the reason, Harry Rowell must have rubbed his hands in expectation as he conveyed his wishes to the battered Buster Ivory. "Buster" he said, "You just leave the ones you boys rode back in Canada, and ship me the rest."

Saddle Development

The northern American broncs were responsible for the development of the undercut or swell fork style of western saddle. The early Texas saddle with a smooth or slick fork (front or pommel) was simply not up to the job handed out to riders on the northern ranges by the big, strong, clumper style horses they had to ride.

Left: Undercut Fork and *Right:* Slick Fork saddles made by the author.
Photos courtesy Jack Drake collection

The development of the undercut fork saddle corresponds with the way the Australian stock saddle evolved from the flat English style brought over by British pioneers. Breaking in mature age station bred horses and brumbies caught from the wild, as well as riding at speed in mountain and scrub country, soon showed up the disadvantages of the flat English saddle. The higher back, deep seat and knee pads came into use over time and gradually the Australian saddle became as we know it today.

Jack Wieneke is credited as the first producer of the Australian stock saddle in its present form with high set knee pads, a four to five inch dip in the seat, and long flaps [5]. John J. Wieneke began making saddles in 1880 at Roma after working as a stockman. He collaborated with the scrub riders and horse breakers of the Brigalow country and the Wieneke style of saddle has survived to this day with only minor modifications. Wieneke later moved to Brisbane and his 'True to label Genuine Wieneke' brand has passed through several companies and is today owned by C.A.Stephan and Co of Albion, Brisbane.

Cover of 1928 Wieneke Catalogue.

Courtesy Ian McLaine

Hamley's Circle H Brand began as a harness making operation in Ripon, Wisconson in 1852 then shifted to Ashton, South Dakota in 1883 and Kendrick, Idaho in 1890, before locating in Pendleton, Oregon in 1905 where it remains to this day[6]. Hamley & Co. has been responsible for many advances in western saddle design. Their Formfitter and Ellensburg trees became the choice of rough string riders on the northern ranges. The Ellensburg was used as the basis for designing the Association Bronc Saddle in 1919 when the need was recognised for a standardised saddle for rodeo events. Modifications to the tree worked out between Hamleys and representatives of Pendleton, Boise, Walla Walla and Cheyenne Rodeo Committees created the saddle adopted for competition world wide.

When Australian Rodeo was formalised with the formation of the Australian Roughriders Association in 1945, the saddle adopted for the buckjumping event was a poly style stock saddle with a fairly flat seat and small 1 ½" (3.8 cm) knee pads. Despite the fact they were 'self emptiers' that gave every advantage to the horse, some tremendous rides were put up in these little flat saddles.

In the late 1950s the ARRA introduced a rig known as an A Saddle which was a hybrid style with an Australian poly style seat and a double rigging. The back girth prevented the saddle slipping forward and did away with the crupper that anchored the saddle to the horse's tail but was often hard to fit in the chute on a touchy horse.

The international saddle was adopted in Australia in the mid 1960s amid a blaze of controversy in the rodeo world. Australia was the last country to adopt its use, from the four major rodeo nations – the U.S.A., Canada, Australia and New Zealand.

Genuine Wieneke 'Mitchell Break' saddle restored by the author for Ian McLaine, great grandson of John J.Wieneke.

Photo courtesy Jack Drake collection

An unknown rider at the 1947 Warwick rodeo obligingly vacates his seat to give a good view of the Davidson and Smith, Australia's first official buckjumping saddle.

Photo from *Warwick Daily News*, 1947. Courtesy P.Poole

A great try but this rider was well above the 'A' Saddle just before he left Greg Canavan's mighty bucking horse *Golden Guineas* at the 1963 Warwick Rodeo.

Photo from *Warwick Daily News*, 1963. Courtesy P.Poole

Australia's Greatest Buckjumpers

Dargin's Grey and Breaker Morant

On a chilly, mist shrouded morning in 1886 on W. Tindall's *Capertree* property near Lithgow, New South Wales, a bucking phenomenon was foaled.

Misty was the name bestowed on the tiny, grey colt but as the baby grew, the name reminiscent of soft, cool weather became less and less appropriate. By the time Mr. Tindall's TJ brand was applied to his near shoulder and he was gelded in preparation for a life of useful service to man, *Misty* was already showing a marked objection to his planned future. He wasn't big, but he was a fiery piece of work and when Tindall sold him to Mr. W. Gardiner of *Wallerawang*, he probably considered himself well rid of the rebellious grey.

All attempts to break-in the nuggetty gelding were met with spirited resistance, *Misty* seeming to feel his life's work should not involve hacking or harness. Eventually a noted horseman named Arthur Dargin who had been searching for an outlaw worthy of his talents, heard about the nemesis of the Wallerawang breakers and bought the horse whose second owner was probably as glad to see the back of him as his breeder had been.

Sometime before the grey's maiden voyage as an exhibition buckjumper, an old Cobb and Co driver told of a meeting between *Misty* and the laureate of Australian roughriders, Lance Skuthorpe. The eye witness account was given to Athol Smith, a friend of historian Jim McJannett, in the 1940s.

Around 1890 when *Misty* was a four year old and proving to be a handful, he was reputed to have thrown Lance very convincingly when Skuthorpe mounted him in a paddock in the Lithgow area. Lance Skuthorpe's stock in trade was riding buckjumpers, and he would undoubtedly have kept it quiet if one ditched him. There is no way to authenticate the incident, but if it happened, it gives the lie to the legend that Skuthorpe was never tossed.

Arthur Dargin decided to debut his new buckjumper at the Lithgow Show. His search for a horse he could not ride came to an abrupt end when *Misty* now renamed *Dargin's Grey* pelted him in short order and nearly tore the place apart. The just-over-pony sized, pocket rocket bucked right across the top of an elegant carriage and pair and chaos reigned until the horse which perhaps should have been named Dargin's Downfall was captured and removed from the ring.

Arthur Dargin campaigned the grey as an exhibition buckjumper until the early 1900s when he left the buckjump show business. Why Mr. Jones of Lithgow's Zig-Zag Brewery bought the horse remains a mystery. Perhaps he had imbibed a bit too much of his own product when the deal was struck, but if *Dargin's Grey* had ever finished up in the shafts of a brewery cart, it would certainly have been no way to treat good beer.

Martin 'Martini' Breheny had left Skuthorpe's Wild Australia show to form one of his own and had battled on for three years barely making ends meet. Breheny was in Parramatta when he heard of the scruffy grey that nearly wrecked the Lithgow Show. He approached Jones who allowed him the use of the grey for one night for a £5 fee. Martini was thankful Jones did not demand his fiver in advance as he would have been unable to pay.

Dargin's Grey did his part disposing of his challengers and the show made £10. Jones was delighted in his horse's performance as well as his fee and magnanimously allowed Martini to use the horse again for nothing.

A few days later, Martin Breheny purchased the talented bucking horse for £8 and as they say, the rest is history.

Dargin's Grey walked over much of Eastern Australia with Martini's show flattening would-be challengers from Sydney to Thursday Island, which he visited by boat. He got as far west as Winton and Cloncurry and missed very little of Eastern Queensland and New South Wales during his time as Australia's star buckjumper. Many talented rough stock riders took a mount on *Dargin's Grey* and only a handful were successful.

A star rider of the time, was a part Aborigine from Proserpine in North Queensland named Billy Waite. Waite who was one of Australia's truly gifted roughriders made six tries at the grey before declaring "No one will ever stick to *Dargin*. It's hopeless."

The first rider to beat *Dargin's Grey* was a man who has become an icon in Australian history as a poet, horseman and scapegoat of authority. Harry 'Breaker' Morant was a wild, devil-may-care Englishman who wandered the Outback in the 1880s and 1890s. Morant has inspired several books and his life and eventual death under the guns of a British Firing Squad in the closing stages of the Boer War, is one of Australia's greatest stories.

'The Breaker' who would go anywhere for a crack at a renowned bucker, heard about Billy Waite's comment and laid down his own challenge. "No horse has been born that can't be ridden" stated Morant. "I'll ride the grey fury, that I will. Lay the sugar (cash) down."

The Hawkesbury Show, N.S.W. was the venue in May, 1897. Morant backed himself for big money and some even bigger side bets were laid. Arthur Dargin specified that no water was to be applied to the slick poly saddle (wetting the saddle offered more grip), no monkey grips or pulling straps were allowed and if the horse fell, all bets were off, but if Harry could re-mount in the event of a fall, he was allowed to.

Harry Morant rode *Dargin's Grey* to a standstill at the Hawkesbury Show – the first man to ever conquer him. As an old man in Sydney, Lance Skuthorpe who saw the ride, had this to say: "That grey devil did everything on the face of the earth to unseat The Breaker, but turn inside out, but Morant read the grey's every move a split second before the horse made it. Only the very best can do that. The

Breaker had few peers, if any. It was pure grace. Easy he made it look. It was far from that. It was all style, touch, class. I offered him Exhibition Rider (to travel with Skuthorpe's show) more than once as had others. I offered that day too. 'No Lance' he said, 'I could ride a mite once. Too many broken bones behind me these days'. Broken bones my eye. It was being tied down. Quids in his pockets, Harry was off on the town."

Lance went on to say that the match he witnessed between the grey and The Breaker was without peer, the best he ever saw. High praise from a rider of Skuthorpe's calibre. Lance Skuthorpe gave his description of the horse's action and said that:

> *Dargin's Gray matched Rocky Ned for relentless fury. He was perhaps not as fast as Snips and not as rough as Black Beauty (another noted Bucker), but what he was, was something near the best of all their traits. He was the complete bucking horse. No horse knew more tricks, not even Bobs. The grey firebrand was a horse with brains. Such horses are difficult.*

A writer who didn't bother to sign his work in the *Sydney Bulletin* 15th November 1902 claimed sixteen challengers had tried the horse before Morant that day at the Hawkesbury. This piece was written in the aftermath of Morant's execution on 27th February of the same year and feelings were either very pro or anti.

No showman particularly one as competent as Martin Breheny would have abused his star performer that way and it is fairly obvious that *The Bulletin* correspondent was trying to play down The Breaker's achievements.

The Breaker's life and the way it ended is a tale of action, romance and high adventure. Stories abound of Harry Harboard Morant's dare-devil escapades and one that demonstrates the larrikin streak in this British born Australian, was told by an old northern New South Wales horseman Charlie Farlow, who served as one of Morant's fellow Boer War soldiers. A neighbour of mine who knew him when Charlie was an old man of ninety, told me the story [7].

"Along with other new recruits, The Breaker was part of a group being taught horsemanship by an English Cavalry officer. No doubt Harry and his bush bred mates most of whom had ridden all their lives, were a bit amused when they were led over to a wooden horse to receive instruction in saddling and bridling. Every one of the men were required to place riding equipment on the model in the approved military fashion and when his turn came, Harry stepped forward. The officer was standing near the rear of the facsimile when he handed Morant the bridle. The Breaker took the proffered piece of equipment, swung it overhead and delivered a tremendous blow to the wooden rump near the officer's shoulder.

"What the devil did you do that for man?" spluttered the enraged instructor.

Harry snapped a smart salute to his superior officer with a broad grin and rapped out "Thought the bastard was ready to kick you, Sah!"

The nickname 'Breaker' may not have been hung on Morant just because of his style and class as a buckjump rider, but also from his ability to gentle horses. Before

Harry left England, then known by his real name of Edwin Henry Murrant, he was schooled by one of the nation's best horsemen. George Whyte-Melville was an upper class Englishman who enjoyed a reputation as one of the very best trainers of jumpers in the country.

Harry/Edwin spent his teenage years with Whyte-Melville and was recorded as an outstanding protégé [8]. He also honed his breaking-in and horse education skills at stables and racetracks in Barnet, the principal horse trading centre for London, while he attended nearby Silesa College. Edwin Murrant was a horseman of advanced ability before he came to Australia and adopted the mantle of 'Breaker Morant'.

Morant was reported to be somewhat aghast at first, at the way Australians broke-in horses. Of course a chap straight out from England would have never been in the situation where he was faced with handling big, mature, bush-bred horses that were unhandled and naturally very wild. Morant soon became expert at handling the worst of outback outlaws but much of his old training remained.

Nickavilla Station in North Queensland employed Morant as head stockman in the mid 1880s [9]. A member of the owner's family remembered the horses he broke-in never bucked with him. He said Harry would talk quietly to the youngsters until he broke down their resistance and the animals would eventually nuzzle under his arm. The account also went on to say that the lady of the house had something of a crush on The Breaker and would go misty eyed and describe him as a 'splendid chap'. Morant certainly had the touch with horses and women.

What a bad blue Lord Kitchener and the British Army made believing him to be a 'nobody' who would soon be forgotten, when Morant was executed. The Breaker was a legend in the Australian bush long before he made the fateful decision to fight in the Boer War.

Other reports have Morant handling wild horses with a rag tied on a light stick to get the animal accustomed to being handled without having a man too close for its comfort [10]. The Breaker called this a 'dolly' and today horse whisperers are using a technologically advanced design of Breaker Morant's wattle branch they call a 'carrot stick'.

Breaker Morant like Lance Skuthorpe, had a marvellous rapport with horses. Both men were not only phenomenal roughriders but all-round horsemen of world class. The legends of Breaker Morant and *Dargin's Grey* are inextricably linked.

Only a few riders went the distance on *Dargin* from the time of Morant's triumph until the early 20th Century when age caught up on the grand old campaigner.

Billy Waite who genuinely believed him to be unrideable, had no doubt fallen victim to 'mind over matter' and inspired by The Breaker's Hawkesbury exhibition Waite got the job done at Sunny Corner in 1897 and again at Bathurst in 1898. In 1898 Tom Sloan rode *Dargin* also at Bathurst. 1899 saw a qualified ride by Harry Mantle at Lithgow. Abel Woods rode him twice at Gunnedah in 1902 and Ned Earl got the grey covered at Cairns and Mossman in 1903.

Other known rides but not dates, were put up by Jack Prendergast at Windsor, Willie Schultz at either Ravenswood or Charters Towers, Peter Molloy at Wellington and Paddy Caton at Stanthorpe [11].

After Stanthorpe's Paddy Caton rode *Dargin's Grey* his employer C.F.White of *Pikedale* Station backed him to ride the horse's successor, *Bobs* with big money the next time Martini brought his show to Stanthorpe. Paddy's grandson Barney Belford who is a knowledgeable amateur historian, takes up the story:

> *Old Charlie White never offered Grandad any extra money and didn't even offer him a bonus on his station wages even though he stood to make a lot of money. Paddy was a little cranky about this and everyone who saw the ride reckoned he let himself get thrown. Charlie White who was a real little bloke and really wild at the time, reared himself up to his full height which wasn't much, and told Paddy to "Never show your face on Pikedale again."*

Rocky Ned

Another Australian bucking horse legend was a chestnut foaled in 1911 on *Box Hill* Station, Bingara NSW owned by Mr. Thomas Butler [12]. He broke-in quietly but at some stage (there are several tales of the exact circumstances) something frightened him and *Rocky Ned* as he later came to be known, threw his rider and was never ridden again without violent protest. They broke the chestnut gelding to harness and he worked quite well in that role until Lennon's Circus and Tom Handley's Buckjump Show put on a combined performance in Warialda NSW around 1920. His owners offered him as a challenge horse and *Rocky* did alright unceremoniously dumping the show's top rider. Mrs. Lennon purchased the gelding and some time later when the two shows parted, he passed to Tom Handly for £10.

Tom Handley's Show featured *Rocky Ned* until 1934 when Tom quit the business and the old horse passed to the well known Thorpe McConville whose show was only surpassed in the annals of rope ring history, by that of Lance Skuthorpe.

Ned's style was to kick very high behind and he had a lot of power. The reason he was so hard to ride seems to be the speed and unremitting aggression with which he bucked.

Many claims have been made about who rode *Rocky Ned* but out of the labyrinth of fathers, uncles, grandfathers and others whose adoring descendants indignantly claim the truth of their family legend, only five claims can be authenticated. The issue is also confused by the fact that the famous buckjumper was the third horse to go by the name *Rocky Ned*. A showman known as Bronco George had a big, bay thoroughbred that bucked under that name around 1910 and a piebald horse belonging to another show run by Jim Fisher in 1918 also used the name [13].

However, the *Rocky Ned* that history remembers was first ridden in 1927 when Billy Timmins conquered the chestnut outlaw at Mungindi NSW. He rode the horse until it stopped bucking twice at Mungindi and there are claims which cannot be substantiated, that he also rode him at Moree. However, no doubt exists that Billy Timmins rode *Rocky Ned* twice in the rope ring of Tom Handly's Buckjump Show. Billy's son Wayne 'Stumpy' Timmins went on to become an Australian Champion and one of only four people to win both the Gold Cup Campdraft and the Open Buckjump (bronc ride) at the famous Warwick Rodeo.

One of Australia's truly gifted horsemen Jack Stanton of Scone in NSW rode *Rocky* at Bundarra while travelling as a rider with the show. Jack had tried him dozens of times and the ovation he received when finally successful was one of the high points of an outstanding career in many areas of the horse business. The Stanton family like the Timmins's, produced champion riders in successive generations which would suggest it took horsemanship bred in the bone to conquer the chestnut outlaw.

Gordon Attwater took up the challenge in the Grafton Showgrounds in 1929 for a purse of £200, a good chunk of money in those days. Claims have been made that Attwater had an unfair advantage as he rode in his own saddle (a Kemp Poly) but in those days buckjumping saddles were not standardised and Tom Handly allowed him to use it. It seems the unfair equipment claim may have been sour grapes on the part of lesser riders. From my own experience as a rider and saddle maker, I can guarantee no saddle ever built can make you unthrowable and the only way you can't come out of them, is straight down.

Later that year, Hilton McTaggart of Goondiwindi, Queensland rode the horse at Scone and two years later he tangled with *Ned* again in the Sydney Showgrounds when The Royal Agricultural Society hired Tom Handly's buckjumpers to stage the Australian Championships. Hilton took out the title with *Rocky Ned* as his finals horse. Hilton McTaggart had five sons who followed in his footsteps and became great horsemen winning Australian titles in the Cutting and Rodeo arenas. So once again, it took someone from a family of riders to handle the chestnut.

Jim Caton of Killarney in Queensland was the other rider to go the distance on *Rocky Ned*. He made the ride when *Rocky* had just come back from a spell and some people claimed the horse was too fat to buck at his peak. Anybody who has ever got screwed down on a wild one knows the fat horse can be a lot harder to ride. To keep the saddle on a horse with a lot of condition, it needs to be girthed tighter than normal. This means the rider absorbs more jarring impact as the horse's feet hit the ground. From photographs it can be seen *Rocky* had a good sized wither and Jim may still have been able to girth his saddle slightly looser to reduce the jar. That is if the showman allowed him to saddle the horse himself, which many did not. However, *Rocky Ned* on a bad day was better than ninety nine percent of other buckjumpers on the best day they ever had.

Jim Caton was no bush mug who got lucky [10]. The Caton family were renowned buckjump riders in the southern downs region of Queensland and northern New South Wales. Although they didn't travel a lot, they took on the best of the buckjump showmen's horses and bested them. Jim's cousin Paddy of Stanthorpe, rode *Dargin's*

Grey' and Bill, another cousin, rode Lance Skuthorpe's 'Queenslander' a renowned tent show bucker.

As an old man, Jim Caton was hurt when a bad storm hit Killarney in November 1968. When the wreckage of his hut was lifted off him, family member Barney Belford quotes Jim as saying "Hi fought in World War I and hit was a bastard. Hi fought in World War II and hit wasn't much better. Hi rode Rocky Ned, but this his the worst fight hi hever 'ad."

It is also worth mentioning that when Rocky travelled with Thorpe McConville's Wild Australia Show, Thorpe's son Doug made many exhibition rides on him. Doug McConville always jumped off the horse after a short time and never rode him right out despite the fact he almost certainly could have.

Rocky Ned lived to the ripe old age of thirty-one years and is buried at River Bend on the banks of the Murrumbidgee River near Narrandera in NSW. Many great bucking horses have lived to big ages. Perhaps that reinforces the theory that only the good die young and it certainly takes the credibility out of the argument that rodeo is hard on horses.

Curio

Curio of The Marrabel Rodeo Club in South Australia, was a roan, brumby mare that came out of the desert, bred in hard country on *Macumba* Station once part of the Kidman Empire in the north west of South Australia. In 1945 she began her legend as the star of Marrabel's bucking string, by ironing out Noel Bottom a very competent rider with a string of credits, in three seconds [14].

She is reported to have flattened Les Cowen in 1946, but Les himself claimed in a letter to *Hooves and Horns* Magazine , *Curio* bucked behind the chute and almost went under the saddle rail before she blew back out and threw him [15]. Les contends he taught *Curio* the suck-back that made her nearly unrideable. Les also stated in his letter that Leo Reichstein rode *Curio* to full time, ten seconds in those days, in the bareback ride in 1945. This does not appear in any official records but according to Les, Reichstein rode *Curio* using a bullock riding rope rather than the leather and rawhide 'riggings' bareback riders use today.

1947 saw her first trip as feature horse and Johnny Pierce managed two and a half seconds on her. 1948 saw her put Alan Bennett down in three seconds. In 1949, they played *Curio's* own song composed and sung by Smokey Dawson as 5,000 people watched Dally Holden, the current Australian Champ, ease down on the mare. No respecter of Dally's accolades, the little roan skyed him in four seconds. 1950 – Ray Crawford – 2.5 seconds. 1951 – Johnny Roberts, Queensland Champion – 1.8 seconds. John Cadell lasted just over two seconds in 1952. He was the man who drove her in from *Macumba* Station and an accomplished buckjump rider, but *Curio's* nemesis was approaching.

In 1953 Alan Woods slipped into the little flat Davidson and Smith Australian Roughriders Association 'rodeo special' saddle. He got his feet in the irons, took a firm grip on the bucking rein and nodded for the gate. He rode her through the first two bucks and then ran into the devastating suck back that had put paid to all her other challengers. Alan was in some trouble with a lot of daylight showing between him and the saddle, but Woods was a thinking rider with reflexes like a cat. He regained his seat and rode the mare to the ten second time limit.

A little known piece of history about that famous ride, is that the man who arranged the purchase of *Curio* and some other Marrabel buckjumpers, and who was the pick-up man who lifted Alan Woods off *Curio* after the ten second time limit, was none other than Australia's legendary bushman's outfitter, R.M.Williams [16].

In an interview straight after the ride, it was suggested that Alan fluked it and *Curio* had jumped back under him after he was thrown. Alan Woods was regarded as the best of his time by his peers, and he answered by offering to do it again in 1954 which he did. The roan brumby bucked just as hard next year and there were no accusations of fluking it that time. Woods rode her fair and square for the second time.

Later that day Buddy Gravener drew the mare in the consolation buckjump. Buddy lost a stirrup around the six second mark and grabbed down with both hands [17]. While not a qualified ride and worth nothing in terms of prize money, Gravener was still actually aboard the roan when the whistle blew.

The decision was made to breed from *Curio* to try and retain the awesome bucking ability of this nondescript little mare. Because she was in foal *Curio* did not buck in 1955 but she came back to rip Billy Austin's hopes to bits in 1956. She was in foal again in 1957 and the following year a great all 'round horseman named Alan Henshke took the ride and the roan fell just outside the chute. Henshke was a big powerful man and many who saw it believed he pulled the horse down deliberately. In earlier times many bush horsemen were extremely clever at pulling a bad bucker off its feet and whatever happened that day the thing we know for sure is that Alan Henschke collected the money for riding *Curio* as he was in control when she fell [18].

Four foals were produced which all went on to buck off some riders. She was back in form in 1959 when she disposed of Brian Gill in two seconds and in 1960 once again a hush settled over the Marrabel arena. Noel Toomey was the name drawn to try Australia's greatest feature horse. *Curio* did not disappoint and a few seconds later she did a victory lap with empty saddle and stirrups flying much to the delight of the 4,000 fans who watched her.

She slipped and fell due to recent rain in 1962 giving Doug Edwards a 'no ride'. He too collected the £100. In 1963 the mare came out in the buckjump but was not used as feature horse. She was ridden by Ron Brewer for 81 points which put him in second place to Terry Moody on *Hector* by a point. Her devastating suckback did not appear.

1964 saw another qualified ride by Dick White who scored 78. Despite turning in a good performance, *Curio* appeared a shadow of her former self to those who had seen her buck at her top. So ended the career of *Curio* of Marrabel.

Her first born, *Son of Curio*, made a very convincing job of tossing Bonny Young, one of Australia's very best riders in the 1960s and 1970s. at the 1965 rodeo while his mother grazed in well-earned retirement.

She passed away in her paddock at twenty-eight years of age. A long serving Marrabel Committee man named Jack Michalanny kept *Curio* on his property northeast of Marrabel during her last years. He provided her with fine grazing, water and shade even bringing her bran mashes when age finally caught up with the grand old mare. *Curio* was a legend of the Australian bush and is buried in the rodeo grounds at Marrabel [19].

American Bucking Stars

Steamboat and Others

About 1901 a horse was foaled on the Two Bar Ranch in Wyoming that came to Cheyenne Frontier Days rodeo in 1905. *Steamboat* bucked off all comers and in the ensuing years, became the first bucking horse to gain an international reputation [20]. He threw them all including Hugh Strickland probably the greatest American all round cowboy of the era.

In 1908 at Cheyenne where he began his notorious career, *Steamboat* was ridden by Dick Stanly of Portland, Oregon. *Steamboat* had one more qualified ride made on him in 1913 by Henry Webb and he died the year after from an injury sustained at Salt Lake City.

Hells Angel belonged to stock contractor Everett Colborn and was easily America's top bucker from 1938 to 1941 [21]. He was only ridden by a select few. Fritz Truan (1939 and 1940 world champion) got the blaze-faced demon ridden five times out of seven but Truan was some kind of a bronc rider and the world of rodeo lost a great performer when Fritz Truan died with the American forces in World War II as they took Iwo Jima Hill.

American ranches and rodeos saw many a good bucking horse like *Flaxie, Done Gone, Double Tough, K.C.Roan* and *Tipperary* as well as modern superstars like *Descent* and *Major Reno* but the two that compare with Australia's *Dargin's Grey, Bobs, Rocky Ned* and *Curio* are two big blacks, *Five Minutes to Midnight* and the horse he took his name from, *Midnight*.

Midnight

Midnight, just like his Australian equal, *Rocky Ned*, is the subject of many conflicting tales about his entry into the cowboy conquering business. He is credited on one hand with being a schoolmarm's buggy horse and on the other as an unbreakable outlaw from the word go as well as a few more unlikely tales. The most creditable version is given by Frederick Melton 'Foghorn' Clancy [22].

Clancy was America's number one rodeo announcer from the late 1890s 'till 1947. He was on the scene during *Midnight's* whole career and was a known and respected authority on rodeo.

The big black was foaled in 1914 on the Cottonwood Ranch in Alberta, Canada. He was apparently the result of an accidental mating between a thoroughbred mare and a Morgan percheron cross stallion, a likely mix for a bucking horse.

Midnight was broken in and used for a while as a cowhorse by Jim McNab of McLeod, Canada, but he threw Jim one day when a car backfired near him and his true vocation soon became obvious.

In 1924 he made his arena debut at the Calgary Stampede where he bucked off Cecil Henley, a leading competitor of the times, very convincingly. Jim McNab sold the horse to new rodeo producers Pete Welsh and 'Strawberry' Red Wall, and *Midnight* continued on his merry way tossing all comers at Wall and Welsh's rodeos.

The partnership hit a rough patch financially and when their assets were put up for auction a horse dealer from Toronto named Mitchell bought the stock. Jim Eskew a Wild West Show promoter was showing at the Canadian National Exposition that year – 1928. Eskew was a great promoter whose commanding presence would later give him the honorary title of Colonel Jim. He wasted no time in seeking out the dealer who really did not know *Midnight's* reputation and value. Eskew purchased the black horse and another slightly smaller black who was destined to become *Midnight's* alter ego *Five Minutes to Midnight*. The horses were far too good for the amateur challengers who patronised his Wild West Show and Eskew bought them as an investment with the intention of selling to a major rodeo promoter.

Jim bought the horses to the Southwestern Exposition and Fat Stock Show at Ft.Worth, Texas in 1929 where he sold them to Verne Elliott and Eddie McCarty who were supplying the bucking stock.

Midnight proved a good buy for McCarty and Elliott and for the next four years he unloaded everyone who left the chute on him including champions like Earl Thode, Pete Knight, Doff Aber and Paddy Ryan.

Midnight's devastating bucking style was said to be a mixture of savage power and a double kicking action. Cowboys who fell victim to the black terror said when it felt as if he had kicked up behind as far as he could, his hind legs would lash out yet again. This sounds rather like *Bobs* and his buck within a buck, and it is hard not to wonder what the result would have been if Lance Skuthorpe had ever given *Midnight* a try.

Midnight was a very gentle horse as long as you didn't try to ride him. Verne Elliott's wife made a great pet of him and always carried sugar cubes when she visited her husband's star performer.

By 1932 *Midnight* had begun to develop a ringbone condition in his front feet. Like champions in any form of endeavour, he had never spared himself in his efforts to win and the years of pounding had taken their toll. Some accounts say two or three qualified rides were made on *Midnight* when he slowed down. Others including Foghorn Clancy, claim the horse was never ridden. Verne Elliott made the decision to retire him in 1933. It is to Elliott's credit that once the great bronc's performances began to fade, he did not let every hick in the country feed their egos by riding *Midnight*.

One of the top saddle bronc riders of the twenties and thirties, Turk Greenough, took the final mount on *Midnight* as an exhibition ride at the 1933 Cheyenne Frontier Days Rodeo. It was a fitting finale to a wonderful career when Turk hit the arena dirt a few jumps out of the chute.

Midnight retired to Verne Elliott's ranch near Plattville, Colorado where he ran 'till his death at twenty-two years of age, in 1936. The bronc riders who had been speared hell west and crooked from his plunging back over the years were invited to write his epitaph. After lots of deliberation, they came up with this inscription carved on the plaque above *Midnight's* grave at the Elliott Ranch.

> Underneath this sod lies a great bucking hoss.
> There never lived a cowboy he couldn't toss.
> His name was *Midnight*. His coat black as coal.
> If there's a hoss heaven, please God rest his soul.

Five Minutes to Midnight

When Col. Jim Eskew bought *Midnight* and a slightly smaller black horse to Fort Worth in 1929 and sold them to McCarty and Elliott, the smaller horse rejoiced in the name of *Tumbling Mustard*[23].

Pete Welsh and 'Strawberry' Red Wall had acquired the black from the Sarcee Indian Reservation in Canada before they went broke and the horse came into Eskew's hands. Once *Tumbling Mustard* began throwing cowboys at major rodeos people often asked "Is that 'Midnight'?" The answer usually given was "Darn close to it" and when *Midnight* was retired in 1933 *Tumbling Mustard* became *Five Minutes to Midnight*.

Old Five as he became affectionately known by the rodeo riders, was ridden to the bell fourteen times in his 1,200 trips into the arena. He was a thinking sort of horse with a large repertoire of unpredictable tricks that he called on to unseat his riders.

The black horse with one white hind foot, a star and a snip, travelled all over the United States and even had a trip to England. He put everything he had into his own particular equine art form and in a career that spanned twenty years, he cost a lot of cowboys a lot of entry fees.

Verne Elliott pensioned *Five Minutes to Midnight* off in 1946 and when he died a few years later, he was buried next to his namesake *Midnight*.

Who Was the Best?

There is no way the greatest bucking horse, ride or rider of all time can be named. Many great rides have been put up on properties where only a handful of people saw them. Some horses have fired hard once or twice and then thrown the towel in. It mostly comes down to the publicity generated at the time. I would suggest *Rocky Ned* must take the honours as Australia's best. In an estimated 4,000 trips into the arena he is reliably reported to have been ridden twice by Billy Timmins and Hilton McTaggart and once each by Jack Stanton, Gordon Attwater and Jim Caton. To be fair to today's competitors, it must be remembered that in earlier times, a ride was only recognised if the horse was ridden until it stopped bucking. *Rocky Ned* was ridden past 10 seconds on several occasions. *Dargin's Grey* must rate highly but his career did not last as long as *Ned's*. *Curio* only bucked one day a year, but to her credit, every rider who tried her was a champion so the little roan didn't get any easy victories.

Midnight seems to get the nod from the Americans as their all time best and is the only internationally known outlaw which appears to have never been ridden. He must have been one hell of a bucking horse!

If we look at statistics, a black horse named *King of the Ring* would have to rate highly. The big thoroughbred came down to Condamine in Queensland from the Northern Territory with a droving plant in the early 1940s. He was in light condition and a local racehorse trainer liked the look of his stripped down frame. The trainer bought the horse and was given a piece of advice by the droving contractor. "Don't take pity on him and feed him up. He will buck and he can buck." [24].

The horse was put into training in Toowoomba. Against the advice of the drover, he was filled up with oats and corn and in a training run he 'dropped out' (began bucking) and pelted his jockey. A strapper who could ride a bit got on and followed the jockey in fairly short order. It was time to call in the heavies.

The Cook boys of Toowoomba, George, Joe and Jimmy, were good roughriders who ran their own local buckjump show. Jimmy Cook got on the racetrack outlaw and he soon took his lumps along with the jockey and strapper causing the racing fraternity to lose interest in the Territory drover's horse.

The black horse went to the Warwick Rodeo bucking string where he went unridden for two years and was used as a feature horse. Roughrider and bulldogger Danny Mahon of Gowrie Junction near Toowoomba, acquired him for his team of buckjumpers and the big thoroughbred now named *King of the Ring* was ridden by Angus Frame at his home town Chinchilla, in 1945.

The fact that *King* was finally beaten may have prompted Danny Mahon to sell him to Angus along with seventeen other horses Frame bought for his travelling buckjump show. Angus bucked the horse until the following year when he sold him to Jack Gill Senior who travelled the big horse all over the eastern states with Gill's Buckjump Show. In a realistically estimated fifteen hundred trips into the ring, he was ridden to the 10 second time limit only once or twice [25].

It is interesting to note that Alan Woods who beat *Curio* twice in a row and was known as one of the best feature horse riders Australia has seen, tried *King of the Ring* eleven times and was defeated on each occasion [26]. In 1956 *King* died of tetanus from a cut hoof at Mackay, Queensland but the way he was going, he could well have been a rival for *Rocky Ned*.

Non Competitive Champs

In Australia as in America, the list of competent rough riders is a long one. Every district had its champions and only those who took to competitive riding achieved much documented recognition. The most complete list of top Australian bush riders was painstakingly researched by Bruce Simpson and appears in his book *Hell, Highwater and Hard Cases*.

Many great station and ranch riders never took to competition. A person Lance Skuthorpe personally rated as one of Australia's greatest roughriders, was Sir Sidney Kidman 'The Cattle King' [27].

A few times when Skuthorpe's show was in Kapunda, South Australia, Kidman's home town, or in other places Sid happened to be, he contacted Lance with a message like "You can let people know I'll come down and ride a horse as long as we can sling a few bob to the people at the "Inland Mission" or some other pet cause of Kidman's. Sir Sidney gave literally thousands (millions in today's equivalent) to charity and Lance Skuthorpe always played the game.

Once Skuthorpe was saddling a yellow bay horse he called *Fausho Ballagh* when Sid arrived after having agreed to ride at the Gawler Showgrounds. Lance told Sid he intended to ride the horse himself as he was too good for the other riders travelling with the show. Sid Kidman gave a grin, adjusted the stirrup leathers, mounted *Fausho Ballagh* and requested the handlers to release him. He threw the reins loose to give the horse every chance, and belted him with his hat every buck until the yellow bay stopped defeated.

No doubt the advertising value of having the greatest landowner in modern history performing in his show would have paid off handsomely for Lance as well as the 'Cattle King's' designated beneficiary.

Sid Kidman spent his early life on horseback and was never thrown off any horse Lance dished up to him. Skuthorpe knew his crowds too well to try cheating them by giving Sid an easy mark and the millionaire grazier handled his best buckers with style and aplomb.

In the 1940s Cammy Cleary from the Barkly Tablelands in the Northern Territory, was recognised as the best station roughrider of his time [28]. A balance rider who rode with only his toes in the irons, he would often reach down and tickle the flanks of bad buckjumpers to urge them on to greater efforts.

Tom Lloyd was a cousin of the notorious Kelly family and a great horseman in early Victoria [29]. His sons Leo and Tom Junior went on to become roughriders with Thorpe McConville's travelling show.

American Ranch Riders

'The Breaker's' American Counterpart

Harry 'Breaker' Morant was often described as being the only Englishman in this country who could ride as well as the pick of the native born Australians. 'The Breaker' was one of the very best this country has seen and apparently another bronco busting Englishman showed up in America around the same time as Harry Morant was beating Australia's best.

Deer Trail, Colorado organised a big cowboy contest on the 4th July 1899 with matched races, steer roping and all manner of horse back games including a bronc riding competition. The worst buckers from ranches for miles around were assembled, and among the entrants was an English born employee of the Milliron Ranch named Emilnie Gardenshire. You'd have to be able to handle yourself to carry a name like that around, and like Morant in Australia, Emilnie did his British heritage proud.

The rankest horse there was a mad bay named *Blizzard* and no doubt a few of the local cowpokes grinned in anticipation when the blindfold came off and Gardenshire commenced battle. The Englishman held the reins loosely in his left hand and went to work with the whip in his right. The harder *Blizzard* bucked, the more Emilnie Gardenshire plied his riding crop and after some minutes, the defeated outlaw gave it best and his rider loped him 'round in a circle before dismounting.

Gardenshire won a suit of clothes for his effort and the title 'Champion Bronco Buster of the Plains' proving that some 'pommies' can hold their own in any company [30].

A Roughriding Artist and Writer

One of America's great, ranch riders who won his spurs on the big northern broncs, was the cowboy artist and author Will James who wrote the western classic *Smokey the Cowhorse* [31].

Will was born Ernest Dufault of French Canadian parents in Quebec, Canada. In 1907 he headed west as a fifteen year old lad to fulfil his dream of being a cowboy. His talent as an artist and later as a writer brought him into the public eye but Will (Bill to his friends) James as he called himself from the time he left home, earned a reputation amongst his cowboy peers as a double tough, bronc rider and a very competent roper and all 'round cowboy.

One of the classic cowboy songs of all time *'The Strawberry Roan'* written by Curly Fletcher was inspired by a mighty try at an outlaw roan Will made at the Ricky Land and Cattle Co. in Mono County California in 1916.

Will was getting dizzy and the pounding the horse was giving him had made his nose bleed when he was unseated by the 'high dive' celebrated in the last line of the song. The horse continued bucking furiously and the saddle loosened and slipped beneath its belly. Will wanted to save his saddle, so he roped and threw the Strawberry bronc with a forefoot catch. While he tried to pull his saddle from amongst the flashing hooves, the roan proved he wasn't finished with James yet by kicking him in the jaw and knocking the cowboy out cold.

Kitty Gill puts up the winning ride in the Ladies Buckjump at Warwick Rodeo, 1947 on *Badger Mountain*.

Photo courtesy Peter Poole

Roughriding Ladies

In the early days of competitive rodeo, the buckjumping event was one of the first basions of women's liberation. Many bush and ranch bred girls could ride every bit as well as their male counterparts and ladies roughriding contests and female entrants in open events, were common place.

As a wide-eyed seven year old back in 1957 at Outram in the South Island of New Zealand, I saw my first ever rodeo. It was a fairly primitive affair with two shotgun chutes and bush rail yards in the corner of a grass paddock. The Calgary Stampede couldn't have made a bigger impression on me. Riding wild buckjumpers! That was living!

The open buckjump that day was taken out by a female rider. Katherine Bell nee Robinson, daughter of Allan 'Bull Tosser' Robinson of *Rankleburn* Station at Tapanui, made the winning ride on a fast spinning grey that got it all done right in front of the big wooden crushes (chutes).

The saddle ride relied more on skill than strength and was the favoured rough stock event for women. Ned Kelly's sister Kate, was said to have travelled with a wild west show as a lady buckjump rider in the 1880s, and this is highly believable as the Kelly family and their cousins the Lloyds were noted riders [31].

When rodeos or bushman's carnivals as they were first known, became organised in Australia, ladies buckjumping events were a common addition to the program. Kitty Gill of the famous buckjump show and rodeo family was a well known rider as were Nora Holden and Gwen Winter who was tragically killed camp drafting at Tenterfield in the mid 1950s.

In the last two years, female roughriders have returned to Australian rodeo with Warwick again running a ladies' buckjump.

The United States had lady riders like Lucille Mulhall said to be America's greatest cowgirl who once amazed President Teddy Roosevelt by roping a coyote. Other notable lady riders were Alice Greenough, Praire Rose Henderson and Rose Smith [33].

America has run all girl rodeos for many years and today there are many top female bull and bronc riders in the United States of America and Canada.

Rose Smith on the unfairly named *Easy Money* at St.Joseph, Missouri, 1920. Note the dress of the spectators – not much Hollywood influence around in those days!

Photo courtesy John A. Stryker)

Today's Champions

The list of outlaw horses is a long one. Many tough customers came and went with scarcely a ripple to mark their passing. It has been quite a while since a really legendary bucking horse has surfaced. The unrideable animals in rodeo these days mostly come from the ranks of the bucking bulls and bull riding has tremendous spectator appeal.

Bull and bareback bronc riding are contests that had their beginnings as novelty events whereas the saddle bronc event as it is now known world wide, began from the confrontations between determined riders and cantankerous horses in open grazing situations around the world.

I can almost hear the old blokes squarking as I write this, but the competition bronc riders of today are more stylish and just as hard to throw as yesterday's heroes. Modern rodeo cowboys have the advantage of the international contest saddle which suits the straight ahead jumping, kicking style of horse. Horses like this are easier to find a rhythm with and can be spurred from the shoulder to the back of the saddle each jump.

Old fashioned American bronc riders used a saddle with what was known as a centre fire rig with only one girth or cinch as they called it. Like the Australian poly saddle, this allowed more freedom of bucking style than today's double girthed model. Wild horses that stayed in a small area near the chutes or snubbing post, were much more prevalent in those days.

These horses were hard to get in time with, and as many rodeo horses were run in wild back then, the old timers didn't know just how or even if, the horse would buck. Nowadays broncs are bred for the job, and all rodeo horses are supplied by stock contractors. This means the rider knows what the horse will do and can plan the ride accordingly.

However, these days to stay on and spur if you can, is not enough. A rider must qualify by holding his feet over the horse's shoulder in the mark out position on the first buck out of the chute then lift on the head rein and keep a rhythmic spurring action up for the duration of the ride.

There are few surprises among today's rodeo horses but the big powerful, grain fed, saddle broncs found at the top rodeos are very strong to ride and the standard of competition is higher than it has ever been.

Every so often conversation 'round campfires and public bars throws up an individual who makes the claim "I was never thrown". If those big talkers only knew that making that statement proves beyond any shadow of a doubt that there were a hell of a lot of buckjumpers they never rode.

Mind Over Matter

A final point worth considering is that once a horse had achieved a big reputation, many challengers were thrown by the rep, not the horse. If a bush kid had girded up his loins for a try at *Rocky Ned* or one of his contemporaries, it would be very hard to block out the image of the impressive string of victories racked up by the horse. There have been cases where name buckers were ridden by people who knew nothing of their reputation, or had reason to believe they had an advantage over other riders.

At Madison Square Garden in the 1940s which enjoyed the reputation of being America's premier rodeo back then, a funny thing happened that hardly anyone knew a thing about. One of the top cowboys of his time whom we will call Fred, had drawn *Five Minutes to Midnight* in a round of the bronc riding. Fred was talking to a fellow rider before the contest. "I can't ride this horse" he said "he's thrown me every time I got on him. He's too good for me."

Fred's pal winked at him and said, "I'll show you a trick". He disappeared and returned a short time later with a little block of wood about two and a half centimetres thick by five centimetres square.

"We'll put this under the front cinch before we pull it tight" he explained to the intrigued Fred. "It'll dig into his brisket and he'll buck alright but it'll soften him up enough for you to get him ridden."

They saddled the horse and Fred got on *Five Minutes to Midnight* armed with the thought of the secret weapon under his front cinch. He nodded, the horse came out in its usual spectacular way and Fred rode him to the whistle in impressive style, winning the round. He returned to the chutes beaming and whispered conspiratorially to his partner-in-crime, "That's a sure enough way to get tough horses rode."

Fred's mate grinned at him and pointed at the ground inside the chute Fred and *Old Five* had just vacated. There lying in the dirt was a little block of wood about two and a half centimetres thick by five centimetres square.

THE WESTERN MYTH

*It was told me by a bushman bald and bent and very old
Upon the road to Poolyerleg and here's the tale he told.
'Twould seem absurd to doubt his word so honest he appeared
and as he spoke the sou' west wind toyed gently with his beard.*

C.J. Dennis

Western Superheroes in America

The folklore of North America created two great, fictitious, fantasy heroes who have come to embody the tall tale and essential 'bigness' of the Wild West.

Paul Bunyan was the creation of the logging industry in the northern states. He was said to roam the timber country, double-edged axe on his shoulder in company with a blue ox named 'Babe'.

Bunyan was there to break log jams, punish evil doers and generally act the part of an early times superman. He often paused in his travels to lend a hand to weary timber men by stacking up a month or two's worth of logs with a few well placed swings of his double bladed axe. Twenty trees on the forward stroke and another ten with the back-swing was child's play to this larger than life lumberjack.

The south-western states were home to 'Pecos Bill' who fell out of his parents' wagon and was raised by the 'Cuyoats' (Coyotes) on the 'lone prairee'. 'Pecos Bill' became the quintessential cowboy hero, a giant of a man who roped tornadoes and rode them to a standstill, built mountain ranges and dug the Rio Grande River all as asides to his main business of *"huntin' down Ornery Critters"* and *"pertectin' widders and orphans"*.

Australia's 'Crooked Mick' of the Speewah

The Australian equivalent of Paul Bunyan and 'Pecos Bill' was one 'Crooked Mick' who even had his own fantasy realm, The *Speewah*[1].

In a land where properties run into thousands of square kilometres, '*Speewah* Station' was *big*, excelled only by its neighbour (at the back, of course) '*Big Burrawong*'. It was a big country where shearing teams had twenty cooks cooking for the cooks and the tea and sugar were put in billies by boat.

Speewah trees were fitted with hinges so they could be laid down to let the sun past when it came by once a week. Horses bucked so hard cannons had been set up near the breaking-in yards so corn beef and damper could be shot up to thrown riders to prevent death by starvation on the way down.

'Crooked Mick' strode through his fanciful kingdom mustering, shearing, timber cutting and attending to all bush work with blinding speed and deadly accuracy. While droving sheep one day, he was caught in darkness that lasted the standard Biblical forty days and forty nights (a touch long even for The *Speewah* where nights usually lasted only seven days). When daylight finally returned, Mick found he had inadvertently swung the mob up a hollow log and was dismayed to find he'd lost 10,000 head up a branch.

'Crooked Mick's' only equal, in fact his superior at working, riding, fighting, dancing, drinking and anything else you care to mention, was 'Big Barnett' from '*Big Burrawong*'. The two met in a shearing contest once and when the final bell rang, 'Big Barnett' was shearing so fast it took him twenty sheep to slow down and stop. Just the same, he only beat Mick by a few and 'Big Barnett' complimented 'Crooked Mick' by saying he was a nice little feller even if he wasn't much of a shearer.

'Crooked Mick', like 'Pecos Bill' and Paul Bunyan, were the creations of people without books and often illiterate, who while away what leisure time they had with fanciful and improbable tales.

There is duplication between countries and some of Mick's exploits have been credited to Paul and 'Pecos'. It is more than likely the same sort of stories were credited to old world, folk heroes in even earlier times.

The *Speewah* was celebrated in verse by some anonymous scribe.

> *I had a relation on Speewah Station. He told me what he'd seen there.*
> *I know it sounds a bit of a lie, but then I haven't been there.*
>
> *It's big and large and wide and huge and long and deep, this station;*
> *It's twice as long and thrice as wide as many a bigger nation.*
> *The men that ride are the station's pride and thousands of miles they go;*
> *They can ride for years in sweat and tears and never the whole run know.*
>
> *One mighty man so big and tall, four feet between the eyes,*
> *Stayed twenty years upon his horse until he got a rise.*
> *The horses there grow wire, not hair, and buck so wide and high;*
> *The riders all, if they chance to fall, are fed with shotgun pie.*
>
> *So far it is from towns and trams and news and stress and strain*
> *The chaps there never see a sea or ship or church or train;*
> *And the news they get on the wireless set is three days old or more,*
> *For as they say, it's a long, long way to the set on the old Speewah.*

The hills are high and strong and grim and rocky at the peak.
We had to stoop for the sun to pass, as it did full once a week.
For the nights we had, it sounds quite mad, were seven days in length:
But we slept like logs — men, sheep and dogs, - and sleeping gave us strength.

The shearing shed was made of iron, and brick and wood and stone;
It stretched for miles along the creek, its length was never known.
Each catching pen for the shearing men was big beyond all reason —
I lost a pup while penning up; we found his bones next season.

The boss, on his motor bike of course, rushed up and down the board;
And the fleecies, when he motored by, they laughed with one accord.
For well they knew, that cheeky crew, that he wouldn't be back for days;
They could laugh and play right through the day and then lie back and laze.

The ringer was hard and strong and grim and wonderful with the sheep;
His feet were twenty yards apart, his pockets three feet deep.
He could spit a mile without a smile, and that can be done by few;
He rolled his fags with wool and dags, and shaved with a blowlamp, too.

So tough were we, so iron hard, so steel-lined and he-mannish,
For morning tea each man would make a big merino vanish.
The bones we'd crunch to finish lunch, nothing would be over:
The horns, the hoof, some tin from the roof, and we were all in clover.

The tea was brown and sweet and thick and full of stones and tannin;
And once, long, long and long ago, they brewed it with a man in.
By a quaint old bird the tea was stirred in a little rowing boat;
He'd get it sweet, put up a sheet, and with the wind he'd float.

This yard is tall, no doubt at all, and very this-and-thatish,
It's full of rough and ready facts, not scented or top-hattish;
But if you go you ought to know you really must be tough. That's far enough:
But if you go you ought to know you really must be tough.

The ending seems a bit out of metre with the rest of the poem I know, but only a fool or a very, very tough man would argue with anything off The *Speewah*. I claim to be neither.

Mysterious Monsters

Mysterious monsters form part of the folklore of Australia and America with 'Bigfoot' and 'The Sasquatch' constantly cropping up in the U.S.A. and Canada.

The huge apelike creatures have their southern counterpart in 'The Yowie' and reported sightings have occurred to this day, but still no concrete evidence has emerged about the existence of large, upright walking primates in the wilder regions of both continents.

Australia also has 'The Bunyip', a monster of Aboriginal legend and reportedly sighted by some white people in the early days of settlement. Extinctions of species happen constantly and it is reasonable to suppose 'The Bunyip' did actually exist but was on the verge of extinction at the time of white settlement. The more believable sightings describe a seal-like creature inhabiting lakes, billabongs and large waterholes in rivers.

In earlier times, seals were much more prevalent in Australian coastal waters and have been seen a long distance up rivers. In flood times, seals could have been stranded in billabongs. One was actually shot in a lagoon near Conargo, N.S.W. over a hundred years ago. It was stuffed and hung over the fireplace in The Conargo Hotel for many years [2].

It is interesting to note that 'Bunyip' sightings which reported roaring voices, blazing eyes and teeth like crosscut saws, seem to have been exclusively seen by people returning from bush shanties and grog shops but never by those heading towards them.

In recent times this trend was repeated in a book the author of which shall remain nameless, on strange creatures and mythical monsters in Australia.

The number of 'Yowie' sightings alluded to in The Northern Rivers area of New South Wales seemed to outstrip those from the rest of the country. Could this be because of the flourishing drug culture in those parts or am I making unfair assumptions and is North Eastern N.S.W. really the centre of a flourishing 'Yowie' habitat?

Myth –v- Reality

If not as flamboyantly fanciful as the tales of 'Crooked Mick, Paul Bunyan and 'Pecos Bill' or as unproven as the rumours of unknown creatures, the Western Myth is certainly alive and well.

Many dinky di, Aussie bushmen are supremely critical of the cowboy persona. At rodeo time in Australian bush towns until fairly recently, ringers from the stations often had an active dislike for the travelling professional riders who returned their hostility with similar vigour. This may have had something to do with the rodeo boys' popularity with local girls, but it mostly sprang from the working man's contempt for the flashily dressed showman.

The station boys reckoned, and rightly so in some cases, that travelling roughriders were 'Show Ponies' who had learned the tricks of riding bucking horses and bulls, roping and steer wrestling, but would be less than useless on a hard day's mustering.

Many, if not most rodeo hands came from rural backgrounds in earlier times and were as capable as any in a station situation. They preferred the carefree existence of the circuit however. Despite the fact that rodeo in America, which took its name from the Spanish word for round up, and stockman's contests, known as Bushman's Carnivals 'Down Under', developed simultaneously, America was credited as the home of Rodeo.

The flair and pageantry of the American Shows as well as big prize money, turned the heads of Aussie boys and they were soon using the costume and equipment of the overseas stars. This annoyed many people who felt Australia's cattle country heritage was being overlooked in favour of America's.

Many Australians who speak scathingly about 'bloody Yankee Imitators' would be surprised to know that the same sort of attitude is very prevalent among ranch bred Americans.

I was at a horse show in the 1980s when a Quarter-horse event was being run. It was a Western Horsemanship Class and contestants were required to use American style equipment and dress Western for the contest which originated in America, the home of the Quarter-horse.

A couple of weather-beaten old blokes in flat Akubra hats and Moleskin trousers were leaning on the rail near me watching a contestant who had been asked by the judge to dismount and mount his horse. When mounting the competitor stood beside the horse's belly facing forward and swung up from that position leaving the reins hanging loose. "Look at that bloody galah" I heard one of the old fellows say. "He'll get himself cow kicked or that horse'll clear out on him."

Anyone who learned to ride in the Australian bush was taught to rein the horse up so the rein on the rider's side was tighter than the off side, and put their foot in the stirrup from beside the horse's shoulder. These two actions insured if the horse moved, it would come towards the rider and help him into the saddle rather than turn away and leave him behind. Should the horse be prone to kicking, it could not reach forward with a hind leg and hit the rider.

I was mentally agreeing with the spectator's comment when my kids put the hard word on me for an ice cream. We walked off to get them and I ran into an American friend of mine who had migrated to Australia some years before. Chip came from a ranching family and his parents were visiting. "Come and meet my Dad" said Chip.

We walked over to a leathery, old bloke in jeans and a western, straw hat who was leaning on the rail. As we approached, he turned around and pointed to another competitor in the act of mounting. "Look at that damn fool" he drawled. "He'll either get himself kicked or that horse'll run away with him."

The act of mounting a horse from anywhere but in front of the girth was unheard of in American ranch country in earlier times, just the same as it was in Australia until the Western movie industry got hold of the cowboy legend.

The act of mounting from behind the girth started because movie makers discovered if they filmed a cowboy getting on the correct way, they could not avoid getting Roy Roger's or Gene Autrey's or *Trigger's* or *Champ's* bum in the middle of the picture. Most undignified, and not at all proper behaviour for movie stars, human or equine.

The movie-makers always had gentle horses for their actors, so it became acceptable to mount from the back and the practise spread to the Western Show Ring.

Another thing we should not forget is that English riders mount from more or less the Hollywood position and also put themselves in danger and off balance by holding the back of the saddle and having to change their grip to the pommel or front, mid mount. Any ringer who got on a horse that way would be laughed out of the stock camp. Think about it though. Australians' riding style started in Britain. It was adapted to fit conditions in just the same way as the Spanish way was adapted in America. Ringers and cowboys learned a set of rules designed to handle horses which could be unpredictable.

Riders in equine cultures who expected others to do their training for them, did not see the need to protect themselves so clearly and many riding accidents today could be avoided by the application of a little cattle country nous.

Old time American ranchers and cowboys were no different from dinkum Aussie bushmen. They felt just as slighted by the treatment their hard learned traditions were given by the movie and advertising P.R. machine as some Australians do by the adoption of modern American trends in the horse and cattle industry and its associated sports.

Look at old pictures of American working cowboys and they will not look that different from shots of an Australian stockcamp of a similar time.

Riders on the northern ranges wore angora chaps known as 'woollies' for protection from the cold. Cowboys in the south wore batwing or shotgun chaps to protect them from the thorny brush of those regions. In open temperate areas, chaps were often not worn at all. Bits and spurs were utilitarian. Hats came in all shapes and sizes.

There have always been flashy dressers in both countries but the run of the mill rural worker dressed for comfort and practicality. For every American cowboy who ordered a cowhide vest or silver mounted spurs from Sears and Robuck or Montgomery Wards Mail Order Stores, there was a flash Australian ringer writing to R.M.Williams for a black or red shiny satin shirt or a king size rolled brim sombrero style hat.

Fashions come and go and no style of dress or equipment, other than the basic saddlery of the country, can be held up as the true costume of the stockman or cowboy. I notice letters in a few rural publications lately by a woman who is crusading to reinstate her idea of Australian bushman's dress and get rid of American influences. 'Bring back jodhpurs and the cabbage tree hat' she brays in print.

The only time jodhpurs were popular with stockmen was after the Boer and First World Wars when army surplus riding breeches were available very cheaply. Once that supply ran out stockmen went back to beaver-moles and gaberdine riding pants. Jodhpurs were only popular in Queensland and New South Wales. South and Western Australian and Northern Territory ringers wouldn't be seen dead in them and labelled them 'shit catchers'.

The cabbage tree hat was developed in the early days because it was woven from local materials (the cabbage palm) and lost favour except in the hot northern areas, once felt headwear became freely available. It would be a bit hard to bring them back as I doubt if a genuine 'cabbage tree' could be found outside a museum today but Panama and western straw styles are popular in both countries.

In early Australia, stockmen were much nattier dressers than in recent times. Bright blue and red Crimea shirts were all the rage over which an open vest or 'weskit' (waistcoat) was worn. A heavy, leather belt, with pouches for knife, pocket watch, matches, tobacco tin and pistol (yes, pistol - revolvers known as 'squirts' were standard equipment for ringers well into the 20th Century and are still carried today where bull catchers capture wild cattle). Belts were worn slung over the hips of skin tight, white moleskin trousers. Knee high riding boots with high, larrikin heels were the fashion, or elastic sides of the same style worn with leggings or gaiters, as they were known then, and cabbage tree or slouch hats. The outfit was often topped off with a colourful sash at the waist and sometimes a sweatrag (neckerchief or bandanna as it was known in the States). An outfit like that could almost have come out of Hollywood, couldn't it?

People in all walks of life are as fashion conscious as they have ever been, and a style of dress with partly American origins is popular among bush people at the moment. However, they wear it with a distinctly Australian flair. Many wear their hat brims turned down all 'round in what is accepted as the Australian way. Lots of Australians in earlier times and today, roll the sides of their hats up in what has come to be recognised as The American style. This was done when the hat aged and became floppy. An old hat turned down all around could fold over a rider's eyes when moving fast – a dangerous situation particularly when riding in scrub and that is exactly the reason why rolled sides developed in America. A rolled brim is aerodynamically sound for a horseman. It channels the air flow and holds the hat on.

Big trophy buckles are common today because many equine sports award them to winners, but they are often on an Aussie style belt with a built in knife pouch set along the belt to prevent injury in a horse fall. In more recent times in America, the knife pouch has become popular amongst cowboys because it is difficult to get anything out of a pocket while riding a horse.

Jeans are probably the most common trouser style in the bush today but this is not confined to this country. Jeans have been popular world wide since the 1950s. The top brands of cowboy cut denims are very comfortable riding pants being specially tailored for horsemen and women.

The Australian elastic sided boot is still the most popular footwear in the bush although the Cuban heeled, thin soled, stockman style has diminished in favour of the high top, riding boot in recent years, for bush horsemen.

People who ride a lot in their work nowadays, still spend more time on the ground than the old timers did. While the Cuban heel boot was the best and safest for riding, they were murder to walk any distance in and were habitually worn with leggings to ease pressure from the top of the stirrup iron. The top boots available nowadays, have soft cushioned soles and the high top takes the place of leggings making them simpler, more comfortable, all 'round footwear.

By wearing top boots inside the trouser leg, the advantage of leggings which dropped sticks and thrown up dirt out between the legging and the boot, is overcome. In areas where grass seed sticking to pants is a problem, many ringers still use elastic sides and leggings with their trouser legs tucked in.

Saddles have changed a lot in Australia over the years as well as in America. Because Americans use ropes a lot, the strong bullhide covered tree, with a horn to snub the lariat to, became the frame of their saddle of choice. Modern Western saddles are a lot different to the old rigs of the trail driving days, which were little more than a tree with girth riggings and stirrups attached. Styles varied to suit areas and conditions and the region a cowboy hailed from, was obvious by his riding gear.

In early Australia, the flat English style saddle was used and developed into today's stock saddle as described in chapter 8. The modern style of stock saddle with high set knee pads, long flaps and deep seat, really only came into being in the early 1900s. Before then, knee pads were set lower on the flaps and the saddle had a shallower seat. The deep padding on the panel or underside of the saddle, was necessary because horses did big distances in very hot conditions and it was only possible to use a light saddle cloth or the heat created would scald the horse's back. This meant the padding must be in the saddle not under it.

Americans who rode in hot areas like Texas and Arizona solved this problem by using stripped down saddles with an open slot along the back creating a cooling air flow. American saddles have minimal padding usually sheepskin, mounted on boards shaped to a horse's back rather like the military or cavalry saddle. They use heavy saddle cloths to provide softness.

In recent years, the use of portable yards, motorbikes and aircraft for Australian mustering operations and the practice of trucking horses to the work area, has greatly reduced the distance ridden by Australian stockmen and the American board style of tree has gradually replaced the heavily padded panel style which is expensive to maintain.

A hybrid type of saddle with an Australian seat and sheepskin or felt lined skirts and usually the American wide stirrup leathers and fenders, has become very popular.

Nowadays stations keep fewer horses and take more time over their training. Today's riders find the fender saddle makes it easier to get horses responding to leg aids and achieve better control over lateral movement.

It is interesting to note that in recent years Australian stock saddles have become popular in America and some companies are exporting large numbers of them to the States.

The movie *The Man from Snowy River* gave the Australian stockman an international image. If we look at the whole situation objectively, Tom Burlinson in his role as 'The Man' is no more representative of the Aussie cattleman of yesteryear and today, than John Wayne was of his real life counterpart. The image of the stockman being presented to overseas dwellers and urban Australians these days is nothing like the reality.

If an Australian mustering team was suddenly confronted with a stockwhip brandishing, log-jumping lair complete with flying oilskin coat, there is a fair chance the cattle they had in hand would rush. By the same token, working American cowboys would have a stampede on their hands if Clint Eastwood burst out of the brush with pistols blazing.

The movie and promotion industry deals in dreams. Rodeo, campdraft, cutting and horse show champions are heroes to young rural people and they will try to emulate their idol's dress and style. It is no different to city people wearing basketball or rugby league gear. Nothing stays the same and who knows what the fashion will be for future generations.

One thing is certain though. As long as the cattle industry survives, and no matter how much motor bikes, stock transporters and aircrafts are used, the horse will still be needed for the finer points of working cattle.

Apart from the huge popularity of horses for recreational purposes, which is their main function today, as long as people eat beef we will see the stockman or cowboy on horseback as the ultimate symbol of the West and its myths and legends.

Conclusion

So, what do you reckon? Did Australia have a Wild West? Too right it did! How could it not have? All the elements were there and if you have stuck to the saddle with me this far, I hope you have enjoyed reading about it as much as I did putting it together.

Men like Sid Kidman and Charlie Goodnight, D'Arcy Uhr and Wyatt Earp, Paddy Kenniff and Butch Cassidy didn't have a clue they were building a legend, but as long as Western tradition survives, they will never be forgotten.

Like the title of a classic, country song, my heroes have always been cowboys. I take pride in the fact I have ridden and worked with some of the last survivors of the true Outback, open range epics on earth. These tales are only a handful of the stories from Australia's and America's Wild West and I hope they will whet your appetite for more of two continents' fabulous histories.

The last real remnants or those times are slipping away. Modern economic conditions demand streamlined operation but the spirit created by ringers, cowboys, drovers, trail drivers, selectors and homesteaders will ensure that Territory ringers retain and take pride in their bull-tossing skills the same way Nevada buckaroos do in their roping.

We are moving into a time when tradition is becoming less and less relevant to the balance sheet and more and more important to the soul. If legends of the Wild West have little practical bearing on our reality today, they have a huge impact on our identity. Australians and Americans have been shaped by their pioneer odysseys and if I never sell one copy of these two books, it will still be a worthwhile exercise for the spirit.

The similarities of the two Wests have been a hobby-horse of mine for years. Family and friends have never stopped telling me I should write a book, so here it is.

The project has been a lot tougher to stay on top of than some of the buckers I got on in my rodeo days, but now I can walk back to the chutes and say *"Well mates, I've got her wrote!"*

The End

ENDNOTES

VOLUME No. 2
(Chapters 1 to 9)

Chapter 1. Drovers and Trail Drivers:

1. *Memoirs of a Stockman* by Harry H.Peck, Stockland Press Ltd. Vic. 1942
2. *Droving Days* by H.M.Barker. Seal Books.
3. *Taming the North* Hudson Fysh
4. *In the Tracks of Old Bluey* by Bobbie Buchanan. Central Queensland University Press, 1997.
5. *Packhorse and Waterhole* by Gordon Buchanan
6. *Frontier Territory* by Glenville Pike. Glenville Pike FRGSA, Mareeba, North Queensland.
7. *Kings in Grass Castles* by Dame Mary Durack. Constable and Co, 1959.
8. *In the Tracks of Old Bluey* by Bobbie Buchanan C.Q.U.Press
9. *The Murranji Track Ghost* by Darrel Lewis. Article Stockman's Hall of Fame Newsletter, February 1996.
10. *Deep of the Sky* by Tom Ronan. Cassell and Co Ltd. 1962.
11. *Gold to Grass* by Arthur C. Ashwin, Hesperian Press 2002
12. *The Forgotten King* by Jill Bowen. Angus and Robertson, 1987
13. *The Stockmen"* Landsdowne Press
14. *The Longhorns* by J.Frank Dobie, Little Brown and Co. Boston 1941.
15. *Oxford History of American West* Oxford University Press
16. *The Trampling Herd* by Paul I. Wellman. Fireside Press
17. *The Trampling Herd* by Paul I.Wellman
18. *The American West* by Dee Brown. Charles Scribners Sons, New York
19. *The Cattlemen* by Mari Sandoz. Sanders Toronto, 1958.
20. *The Trampling Herd* by Paul I Wellman
21. *The American West* by Dee Brown
22. *In the Tracks of Old Bluey* by Bobbie Buchanan
23. *The Stockman* Landsdowne Press
24. *The Stockman* Landsdowne Press

Endnotes 221

Chapter No. 2 DUFFERS and RUSTLERS:

1. *Waltzing Matilda* by Richard McGoffin. Robert Brown and Associates, 1995.
2. *Frontier Justice* by Tony Roberts, University of Queensland Press, 2005
3. *Gather No Moss* by William Linklater. MacMillan Company of Australia Ltd.
4. *Mighty Men on Horseback* by Tom Ronan. Cassell and Co. Ltd.
5. *The Territory* by Ernestine Hill. Angus and Robertson 1951.
6. *Gather No Moss* by William Linklater. The MacMillan Co. of Australia Ltd.
7. *Gather No Moss* by William Linklater.
8. *The Big Run* by Jock Makin. Rigby, 1970.
9. *Riders from the Warrego* Warrego and S.W. Qld Historical Society. A collection of papers Vol. 3
10. *And it was not Easy* by G.Stack. Boolarong Publications, 1984.
11. *The Last Showman* by James Oram. Pan MacMillan Australia, 1992
12. *The Australian Yarn* by Ron Edwards Rigby
13. The Tombstone Daily Epitaph, October 27, 1881.
14. *Wyatt Earp. The Life Behind the Legend* by Casy Tefertiller. John Wiley and Sons Inc. 1997.
15. *The Trampling Herd* by Paul I.Wellman. Fireside Press

Chapter 3. BLACK, RED and WHITE WARS:

1. *Six Australian Battlefields* by Al Grasby and Marji Hill. Angus and Robertson, 1988
2. *The Australian Frontier Wars* 1788 – 1838 by John Connor. University of N.S.W. Press
3. *When the Sky Fell down* by Keith Willey, William Collins Pty Ltd, Sydney 1979
4. *Six Australian Battlefields* by Al Grasby and Marji Hill
5. *The Australian Frontier Wars* 1788 – 1838 by John Connor
6. *Blood on the Wattle* by Bruce Elder. National Book Distributors 1988
7. *The Australian Frontier Wars* 1788 – 1838 by John Connor
8. *The Australian Frontier Wars* 1788 – 1838 by John Connor
9. *Blood on the Wattle* by Bruce Elder
10. *World Book Encyclopaedia* Field Enterprises Educational Corporation.
11. *Australian Encyclopaedia* Angus and Robertson, Sydney 1958
12. *Massacre: Myall Creek Revisited* by Russ Blanch. Grah Jean Books 2000
13. *Blood on the Wattle* by Bruce Elder
14. *Baal Belbora* by Geoffrey Blomfield. Colonial Research Society 1981
15. *A Nest of Hornets* by Gorden Reid. Melbourne Oxford University Press 1982.
16. *Pioneering into the Future. A History of Nanango Shire* by Dr. Judith A.Grimes. Wise Owl Research Publishers 1998. ISBN 0958547 14 9
17. *In the Shade of the Bunyas. A History of the Maidenwell-Wengenville District* 1882 – 1982 Maidenwell Centenary Committee printed by Cranbrook Press 1982.
18. *A Nest of Hornets* by Gordon Reid

19. *The Rifle and the Spear* C.L:ack and H.Stafford, 1964 Fortitude Press, Brisbane
20. *A Nest of Hornets* by Gordon Reid
21. *A Nest of Hornets* by Gordon Reid
22. *A Nest of Hornets* by Gordon Reid
23. *Cullin-la-Ringo* by Les Perrin 1998 ISBN 0 646 36033 7
24. Robert Bond, Aboriginal Elder and Historian, Cherbourg Aboriginal Settlement, Queensland. Interviewed 29th November, 2004.
25. *Landscape of Change. A History of the South Burnett Vol.* 1 by Dr. Tony Matthews.
26. *A Nest of Hornets* by Gordon Reid
27. *A Nest of Hornets* by Gordon Reid
28. *A Nest of Hornets* by Gordon Reid
29. *The Kalkadoons, a Study of an Aboriginal Tribe on the Queensland Frontier* by Robert E.M. Armstrong
30. *Taming the North* by Hudson Fysh. Angus and Robertson, Sydney, 1933
31. *Six Australian Battlefields* by Al Grasby and Margi Hill
32. *The Gatton Murders* by Stephanie Bennett. Pan Macmillan Australia
33. *Baal Belbora, The End of the Dancing* by Geoffrey Blomfield
34. *The Rifle and the Spear* by C.Lack and H.Stafford.
35. *Royal Historical Society of Queensland Journal, May 2003.* 12th September, 1843 The Battle of One Tree Hill – A turning point in the conquest of Moreton Bay by Frank Uhr.
36. Article from *The Gympie Times* Friday June 3rd, 1994 supplied by Peter Hiendorf, Brisbane. Skyring descendant.
37. *Blood on the Wattle* by Bruce Elder
38. *Frontier* Episode 3 ABC TV Documentaries 1997
39. *Blood on the Wattle* by Bruce Elder
40. *Blood on the Wattle* by Bruce Elder
41. *Oxford History of the American West* Oxford University Press, New York
42. *Australian Encyclopaedia* Angus and Robertson 1958
43. *Bury My Heart at Wounded Knee* by Dee Brown
44. *The Buffalo Hunters* by Mari Sandoz
45. *Bury My Heart at Wounded Knee* by Dee Brown
46. *World Book Encyclopaedia: Field Enterprises Education Corp.*
47. *Bury My Heart at Wounded Knee* by Dee Brown. Pan Books, 1972.
48. *World Book Encyclopaedia: Indian Chiefs.* Field Enterprises Educational Corp. 1971 USA.
49. *American Indian Genealogy* www.accessgenealogy.com
50. *Bury My Heart at Wounded Knee* by Dee Brown
51. *Bury My Heart at Wounded Knee* by Dee Brown
52. Jim McJannett personal collection.
53. *Great Plains* by Ian Frazier. Farrar Straus and Giroux Inc., New York, 1989.
54. *Chasing the Rainbow* by Glenville Pike
55. *Crazy Horse* by Mari Sandoz. J.S.Reginald Saunders Publishers, Toronto, 1964. Library of Congress No. 64 19069

Endnotes

56. *Bury My Heart at Wounded Knee* by Dee Brown
57. *Great Plains* by Ian Frazier
58. *Indigenous peoples Literature* NativeWire.Indians.orgHome.
59. *40 Years of Gatherings* by Spike Van Cleve, Lowell Press, 1988
60. *Rattlesnakes* by J.Frank Dobie University of Texas Press 1965
61. *Oxford History of the American West* Oxford University Press, New York
62. *Desperate Men* by James D.Horan. University of Nebraska Press. Bison Books, 1997
63. *The Columbia Electronic Encyclopaedia* 6th Edition 2004 Columbia University Press.
64. Forest Carter, Storyteller in Council to the Cherokee Nation.

Chapter 4. WHITE AGAINST WHITE:

1. *The Years of Liberty. The Story of the Great Irish Rebellion of 1798* by T. Pakenham: Hodder-Stoughton, 1967 London.
2. *Six Australian Battlefields* by Al Grasby and Marji Hill
3. *Six Australian Battlefields* by Al Grasby and Margi Hill
4. *Eureka: Rebellion Beneath the Southern Cross* by G.M. Gold. Rigby 1977
5. *Eureka From the Official Records* The Story of the Ballarat Riots of 1854 and the Eureka Stockade, from the Official Documents in the Public Record Office of Victoria. Compiled and edited by Ian MacFarlane
6. *Eureka From the Official Records* Compiled and edited by Ian MacFarlane
7. *Six Australian Battlefields* by Al Grasby and Margi Hill
8. *Eureka from the Official Records* compiled and edited by Ian MacFarlane
9. *Eureka from the Official Records* compiled and edited by Ian MacFarlane
10. *The Shearers War* by Stuart Svensen. University of Queensland Press,
11. *Freedom of Contract* A History of the United Grazier's Association of Queensland. Brisbane 1990.
12. *Where the Dead Men Lie* by Bruce Simpson and Ian Tinney. A.B.C. Books, 2003
13. *Freedom of Contract* by Ruth S. Kerr
14. Interview with Peter Casserly. World Shearing Championships, Toowoomba, Queensland 10th June 2004.
15. *Guiness Book of Records*. Guiness Books, 1987
16. *Freedom of Contract* by Ruth S.Kerr
17. *Waltzing Matilda* by Richard Magoffin. Robert Brown and Associates Qld Pty.Ltd., 1995.
18. *Australian Folklore* by Bill Wannan. Landsdowne Press, 1970.
19. Articles by August Lucanus, *Daily News* Western Australia, 1929
20. *River of Gold* by H.Holthouse. Angus Robertson, Sydney 1967
21. Reminiscences of Harry Harbord and Reginald Spencer Browne *Cairns Post,* 1924
22. *Oxford History of the American West*. Oxford University Press, New York
23. *Roughing It* Mark Twain.
24. *The Mountain Meadows Massacre* Josiah F.Gibbs, Salt Lake Tribune Publishing Co, 1910.

25. *The Cattlemen* Mari Sandoz. Sanders of Toronto Ltd. 1958
26. *Desperate Men* by James D.Horan. University of Nebraska Press. Bison Books, 1997
27. *The Trampling Herd* by Paul I.Wellman. Fireside press. W.Foulsham and Co, London

Chapter 5. WOMEN OF THE WILD WEST:

1. *Blood on the Wattle* by Bruce Elder. National book distributors, Brookvale, 1988
2. Oral history of North Queensland
3. *Deep of the Sky* by Tom Ronan. Cassell and Company Ltd. 1962
4. *The Cowboys* Time Life Books;
   Woman of the Wild West by Dee Brown. Pan Books
5. *The American West* by Dee Brown. Charles Scribner's Sons, 1994
6. *Making Peoples* by James Belich, Penguin Books, New Zealand 1996
7. *Folklore* by Bill Beatty. National Book Distributers and Publishers, 3/2 Aquatic Drive, Frenches Forest NSW 2086
8. *The Rare Sex* by Frederick Folkard. Sydney 1965.
9. *The Warrego and S.W. Queensland Historical Society – a collection of papers on history and other subjects relating to Cunnamulla District* Vol. 1.
10. *World Book Encyclopaedia* Volume 14, Field Enterprises Educational Corp 1970
11. Private research collection of Jim McJannett
12. Paper by Carolyn Laffan Victorian Arts Centre, Performing Arts Museum.
13. *Enter the Colonies Dancing, a History of Dance in Australia.* By Edward H.Pask 1835-1940 (Melbourne: Oxford University Press 1979)
14. *Women of the Wild West* by Dee Brown Pan Books 1958
15. *Eureka. From the Official Records.* The Story of the Ballarat Riots of 1854 and the Eureka Stockade from the official documents in the Public Records Office of Victoria. © State of Victoria
16. Articles by Augustus Lucanus, Daily News, Western Australia 1929
17. Newspapers, Census records, articles in private research collection of Jim McJannett
18. *Queensland Police Gazette* 1878
19. *Frontier Justice* by Tony Roberts, University of Queensland Press, 2005
20. *Northern Territory Times* 2.7.1881; 9.7.1881; 17.12.1881
21. *Frontier Justice* by Tony Roberts, University of Queensland Press, 2005
22. Articles by Augustus Lucanus, *Daily News,* Western Australia, 1929
23. *The Birth of Booroloola* article by C.E.Gaunt, *Northern Standard* 13.10.1931
24. Private research collection of Jim McJannett
25. *The Birth of Booroloola* article by C.E.Gaunt, *Northern Standard* 13.10.1931
26. *The Cruise of the 'Palmerston', Northern Territory Times* and *Gazette* 7.3.1885
27. Articles written by Augustus Lucanus, 1929
28. *Northern Territory Times* 14.3.1885
29. *Northern Territory Shipwrecks* by Tom Lewis
30. *Booroloola, Isolated and Interesting* by J.A.Whitaker, 1985
31. *Western Australian Police Gazette*

Endnotes 225

32. *Murchison Times* and *Day Dawn Gazette* 26.6.1895
33. *Papua Government Gazette*, 1907
34. *The Territory* by Ernestine Hill. Angus and Robertson, 1951
35. Jim McJannett personal collection.
36. *The Lady Bushranger* by Pat Studdy-Clift. Hesperian Press, Western Australia ISBN 0 85905 223 0
37. *The American West* by Dee Brown.
38. *Wyatt Earp, The Life Behind the Legend* Casey Tefertiller. John Wiley and Sons Inc. 1997 USA
39. *The Trampling Herd* by Paul I. Wellman. Fireside Press. W.Foulsham and Co. London
40. *A Dynasty of Western Outlaws* by Paul I. Wellman. University of Nebraska Press, Bison Books 1961.
41. *The Law West of Fort Smith* by Glen Shirley. 1957. Library of Congress No. 57 6193

Chapter 6. STAGECOACH WHIPS, PONY EXPRESS RIDERS, BULLOCKIES and MULE SKINNERS:

1. *Cobb & Co Coaching in Queensland* by Deborah Tranter. Queensland Museum
2. *The Lights of Cobb and Co* by K.A.Austin. Rigby Books, 1967.
3. *The Lights of Cobb & Co* by K.A.Austin
4. *Old Coaching Days in Otago and Southland* by E.M.Lovell-Smith. Christchurch 1932
5. *Old Coaching Days in Otago and Southland* by E.M.Lovell-Smith Lovell-Smith and Venne-Ltd, Christchurch 1931.
6. Personal recollections of J. Drake
7. *Frontier Territory* by Glenville Pike
8. *Fought and Won* by John Lewis. W.K.Thomas and Co "The Register" Office, Printers and Publishers, Adelaide 1922.
9. *The Australian Advertiser Eucla News by telegraph* 18.3.1895
10. Articles by August Lucanus *Daily News*, Western Australia, 1929
11. Private research collection of Jim McJannett
12. *Hell High Water and Hard Cases* by Bruce Simpson. ABC Books, 1999.
13. *Packhorse and Pearling Boat* by Tom Ronan. Cassett Aust. Ltd. 1964
14. *Bush Pilot* by Bob Norman
15. Family History collection Stellamary Matheson Drake.
16. *The Overlanders* by Gary Hogg. Pan Books, London 1961.
17. *The Stockman* Landsdowne Press.
18. *The Stockman* Landsdowne Press.
19. *The Oxford History of the American West*. Oxford University Press
20. *The Oxford History of the American West*
21. *The Oxford History of the American West*
22. *The Overlanders* by Garry Hogg. Pan Books, London 1961.
23. *Riverboats* by Bill Wannan, 1978

24. *Riverboats* Video. Australian Heritage Series. Century pacific Corp Pty. Ltd. 620 High Street, East Kew, Vic 3101.
25. *Vanishing Australians* by George Farwell. Seal Books
26. *Oxford History of the American West*
27. *Oxford History of the American West*
28. *Facts and Feats of the Wild West* by Robin May. Hamlyn Publishing Group.
29. *Oxford History of the American West*
30. *Mark Twain: Roughing It*
31. *Facts and Feats of the Wild West* by Robin May
32. *Wells Fargo* by Dale Robertson, 1975.
33. *Mark Twain: Roughing It*
34. *Cowboys and Indians* by Royal B. Hassrick. Octopus Books, 1976
35. *Wells Fargo* by Dale Robertson

Chapter 7. HORSEBORNE ENDURANCE FEATS:

1. *Buffalo Bill. Last of the Great Scouts* by Helen Whitmore Cody. Kensington Publishing Group, New York 1976
2. *Cossack Gold* by W.L. Lambden-Owen. Hesperian Press WA
3. Article in *The Queensland Times* 22nd September 1980. B. Eriksen
4. *Recollections by Mr. H.P. Somerset* published in an Esk, Queensland newspaper. 18th June 1932.
5. Jim McJannett, personal collection
6. *Wells Fargo* by Dale Robertson, 1975.
7. *Born to Fight* by Neil Speed. The Caps and Flints Press, Melbourne 2002
8. *The Trampling Herd* by Paul I. Wellman. Fireside Press. W. Foulsham and Co. London.

Chapter 8. BUCKJUMPERS and ROUGHRIDERS:

1. *The Horse Tamer* by J. Pollard Rigby Limited 1st pub. 1962 as *The Roughrider* Melbourne Lansdowne.
2. *Australian Cowboys, Roughriders and Rodeos* by Jenny Hicks CQUP 2003
3. *My 50 Years in Rodeo* by Foghorn Clancy. The Naylor Company, 1952. San Antonio, Texas USA
4. *Recollections of B. Ivory* interview by J. Osbourne June 2003.
5. *Catalogue Wieneke* courtesy Ian McLain, great grandson of J. Wieneke.
6. *Western Horseman* magazine May 1992 Vol 57 No.5 *Legends in Leather* page 113 by Phetsy Calderon.
7. Interview with Caville Graeme, 8th December 2004.
8. *Shoot Straight You Bastards* by Nick Bleszynski, Random House, Australia, 2002
9. Phone interview with a member of the Robertson family ex-Nickavilla Station, North Queensland, by Ted Robl
10. Private research notes of Jim McJannett

Endnotes

11. Articles, letters from the private collection of P. Poole, historian Aust.Pro Rodeo Association.
12. Interview with B. Belford June 2003.
13. Magazine article by 'Starlight' of Bellingen 1940s.
14. *Hoofs and Horns* magazine November 1985. *Marrabel Rodeo Golden 50 Years* by D.Tilmouth
15. Letter by Les Cowan in "Hoofs and Horns" Magazine August 1981
16. Interview with John Osborne, 18th February 2004.
17. Interview with Peter Poole, Australian Pro Rodeo Historian 22nd October 2005
18. Interview with Peter Poole, Australian Pro Rodeo Historian 22nd October 2005
19. *Marrabel and District. The Legend of Curio* by Stan Rowell and Fred Hausler, Lutheran Publishing House, 1987
20. *My 50 Years in Rodeo* by Foghorn Clancy
21. *My 50 Years in Rodeo* by Foghorn Clancy
22. *My 50 Years in Rodeo* by Foghorn Clancy
23. *American Hoofs and Horns* magazine September 1948 article by A.M.Hartung supplied from archives ProRodeo Hall of Fame, Colorado Springs, USA.
24. Interview with Angus Frame, 26th October 2004.
25. From the personal memoirs of Angus Frame
26. *Hoofs and Horns* magazine *Rodeo out of the Past* by Greg Russell June 1971.
27. *The Forgotten King* by Jill Bowen. Angus and Robertson, 1987.
28. *Packhorse Drover* by Bruce Simpson ABC Publishing
29. *Hell Highwater and Hard Cases* by Bruce Simpson. ABC Books.
30. Field and Farm July 8th, 1899. Quoted from *Trailing the Cowboy* by Westermeir re: quoted *Cowboys of the America's* by Richard W.Slatta. Yale University Press.
31. *Ride for the High Points* by Jim Bramlett Mountain Press Publishing Co. 1987
32. *Australian Roughriders Cowboys and Rodeos* by Jenny Hicks. Central Queensland University Press.
33. *My 50 Years in Rodeo* by Foghorn Clancy

Chapter 9. THE WESTERN MYTH.

1. *Crooked Mick of the Speerwah* by Bill Wannan.
2. *Australian Folklore* by W.Fearn and B.Wannan. Lansdowne Press, Melbourne

INDEX

A

A-Saddle 190, 191
Abbott
   Capt. Edward 88
   Downing & Co (America) 162
Aber, Doff 201
Abilene, Kans 22, 136
Absaroka (Crow) American Indian 84
Adams and Co, USA 143, 160, 178
Adelaide, SA 3, 14
Adelaide Register, News pictorial 75
Afgans, cameleers (Aust) 157
Ainsworth, John 158
Alamo, The, Tex USA 87
Alberta, Canada 201
Albion, Brisbane, Qld 189
Albury, NSW 3
Alder Gulch Goldfield, USA 22
Alhambra Saloon & Gambling House(America) 136
Alice Downs Station, Aust 100
Alice River, Aust 175
Alice Springs, N.T. 34, 174
Amalgamated Workers Union of Queensland 101
American Civil War 1, 21, 47, 83, 87, 138, 143, 160, 162
American Express 160
American quarterhorse 183, 214
American Rangers Revolver Brigade 94
American War of Independence 52, 87, 89
Amor Push, The (Aust) 101
Amor, John 101
Anderson, George 58
Angoodyea, Aust.Aboriginal 38
Angus, Sheriff, Red 109, 110
Annan River Tin Mining Co (Aust) 176
Annan River, Aust 176
Anthonys Lagoon, NT 154
Apache Indians, America 26, 85
Aplin
   Charlie 156
   George 156
Appomatox, USA 21
Arabanoo, Aust. Aborigine 51
Arafura Sea, off Aust 37
Aramac, Qld 5
Arapaho Indians, America 77, 78, 79
Arcturus Downs Station, Aust 97
Aristocrat, bucking horse 187
Arkansas River, USA 77
Armidale, NSW 135
Armstrong River, N.T. 12

Arnhem Land NT 37
Arrawidgee, bucking horse 187
Arrilalah, Qld 98
Arthur, Governor George 56
Ashton, S.Dak 190
Ashwin, Arthur 14, 15
Association Bronc Saddle 190, 208
Atkins, William 167
Attack Creek, Aust 14, 148
Attwater, Gordon 197, 203
Auburn River, Aust 64
Austin, Billy 199
Australian Labour Federation, The 100, 101
Australian Labour Party 103
Australian Poly Saddle 208
Australian Roughriders Association, A.R.R.A. 190
Australian Steam Navigation Co 171
Australian Stock Saddle 188
Australian stockhorse 183
Australian Workers Union, The 101, 102
Autrey, Gene 215
Averill, Jim 108
Ayrshire Downs Station, Aust 101, 103
Aztec Land and Cattle Co. (America) 111

B

Babe, an ox 210
Badger Mountain, bucking horse 206
Badger, Charlotte 118
Bakery Hill, Ballarat, Vic 93
Ballarat Benovolent Asylum, Vic 147
Ballarat Gold Field, Vic 87, 91, 95, 128, 145
Ballarat Reform League, The 92
Ballarat Times newspaper 92, 128
Balmain, Sydney NSW 126
Balonne River, Aust 61
Ban Ban Springs, Qld 5
Banks, Sir Joseph 53
Bankstown, Sydney NSW 185
Barambah People, Aust Aborigines 66
Barcaldine Downs Station, Aust 98
Barcaldine, Qld 96, 97, 98, 100
Barkly Tableland, N.T. 4, 20, 30, 33, 205
Barnet, England 195
Barrier Track, S.A. 13
Barringun NSW 39
Barrow Creek Police Station, Aust 74

Barrow Creek Telegraph Station (Aust) 148
Barwon River, Aust 151
Bassett, Samp 137
Batavia River, Aust 122
Bates
   'Silent Bob' 145
   John 59
Bathurst, NSW 54, 144, 184, 195
Bathurst Plains, NSW 53
Battle Mountain, Qld 71
Battle of Bathurst 54
Battle of Beechers Island, USA 77
Battle of Lukinville, The 105
Battle of One Tree Hill, Aust 73
Battle of the Hawkesbury, Aust. 52
Battle of the Rosebud, USA 83
Battle of the Washita, America 80
Battle of Vinegar Hill, Aust 87
Baulkham Hills, NSW 89
Baxter Springs, Kans 23
Bay of Islands, NZ 118
Beaconsfield Hotel, Qld 130
Beatty, Peter L. 136
Bedourie oven (Aust) 19, 153
Beecham, Henry 39
Beecher, Leiutenant Fred 78
Beecher's Island, USA 79
Behan, John p 45
Behning, 'Dutchy' 37
Belford, Barney 196, 198
Belle Star 138
Bell, Katherine 207
Benagari Station, Aust 58
Bendemeer, NSW 5
Bendigo, Vic 91
Bennalong, Aust.Aborigine 51
Bennett, Alan 198
Benteen, Capt.Frederick W. 80
Bently, James Francis 92
Beresford, Sub Inspector Marcus-de-la Poer 70
Beringarra, W.Aust. 10, 29
Best, Harold 'Smoked Beef' 132
Bidjigal Aust. Aborigines 49
Bielbah, Aust.Aborigine 60, 66
Big Barnett 211
Big Burrawong, Aust. 210
Big Muddy see Missouri River, USA
Big River Aust Aborigines 56
Big River Catchment (Gwydir) Aust 58
Bigfoot 212
Billy the Kid, William Bonney 44, 114
Bingara, NSW 196
Biniguy Station 59
Bismark Tribune newspaper USA 80
Bitter Creek, bucking horse 187

Index

Black Bart (Charles E.Bolton) 164
Black Beauty, bucking horse 194
Black Hills, (Paha-Sapa), Dak. 83
Black Kettle, American Indian 77
Black Shawl, American Indian 83
Blackall, Qld 100
Blackburn, Mr. Justice 72
Blackland, Tex 187
Blackwattle Bay, Aust. 50
Blake, John 59
Blakeney, Judge 33
Blanchewater Station, Aust. 34
Blaxland, Gregory 53
Blevans
   Andy 113
   Charles 113
   Hampton 112
   Mark 112
   Sam Houston 113
Bligh, Governor 90
Blizzard, bucking horse 205
Blondie, bucking horse 187
Blue Duck (American outlaw) 140
Blue Mountains, NSW 53
Blyth, Alex 61
Blythdale Station, Aust. 61
Bobs, horse 135, 185, 196, 200, 201
Boileau, John 'Long John' 172
Bois d'arc Osage Orange, timber 27
Boise, Idahop 190
Boldrewood, Rolf 33
Bolton, Charles E. (Black Bart) 164
Booroloolooa, NT 130
Borrum, Jim p 25
Bosque Grande, USA 47
Bosque Redondo Reservation, USA 26, 47
Botany Bay, NSW 49
Bottom, Noel 198
Boucaut, Ray 148
Boulia, Qld 69
Bounty, ship 90
Bourke, John 3
Bourke, NSW 159
Bowen Downs, Station 5, 10, 33
Bowes, David 103
Box Hill Station, Aust 196
Bozeman Trail, USA .23, 83
Bracewell, a convict .60
Bradley, William B. 143
Bradshaw, Capt. Jack 35
Brazos River, USA 25
Breheny, Martin 'Martini' 135, 185, 193
Brewarrina, NSW 151
Brewer, Ron p 199
Bridgewater Station, Aust. 69
Brisbane, Qld 40, 60, 144, 168, 171
Brisbane River, Qld 167
Brisbane, Governor Thomas 53
Briscoe, Billy 75
Britton, John F. 143
Brocius, 'Curly Bill' 44
Bronco George 196
Bronte, Tex 186
Brooks, Fred 75
Brown Bess muskets 56

Brown
   Alf 37
   Arapho 111
   Jack 14
   T. 69
   W. 69
Bruett, Dr.William A. 180
Brumby, Aust. wild horse 188
Brunette Downs Station, Aust 33
Buchanan track, Aust 29
Buchanan, Nat 4, 5, 6, 8, 9, 10, 12, 24, 29, 33, 176, 178
Buckingham Downs, Aust 69
Buffalo Bill 82, 165
Buffalo, Wyo 109
Bulimba, Qld 73
Bullamon Station, Aust 97
Bullhead, Sgt. American Indian 82
Bundaberg, Qld 167
Bundarra, NSW 197
Bunnia-Bunnia (Bunya nut grounds) Qld 61
Buntine
   Hugh 117
   Mrs 117
Bunya Mountains, Qld 60
Bunyan, Paul 210
Bunyip 213
Burgett, bucking horse 187
Burke & Wills, exploration party, Aust 68, 156
Burketown, Qld 7, 8, 11, 130
Burks
   Mrs. Amanda 29
   W.F. 29
Burlinson, Tom 218
Burns
   Peter 50
   Robbie 175
Burrundie, N.T. 133
Bushy Park Station, Aust 69
Bushmans Carnivals 214
Butler, Mr. Thomas 196
Butterfield, John 160
Byers, William 45
Byron, Lord 127

C

Caboonbah Station, Aust 168
Caddon, Constable John 172
Cadell, Capt.Charles Francis RN 159
Cadell, John 198
Cairns, Qld 42, 134, 156, 172, 174, 195
Calcutta Warship 89
Calgary Stampede, Canada 182, 201, 206
California Gold Rush (America) 5, 21, 112, 127, 160
California Steam Navigation Co. (USA) 158
Callandoon Station, Aust 61
Calvert River, Aust 8
Camooweal, Qld 4

Camp Robinson, Neb 83
Campbell
   Jim 34
   John 'Tinker' 73
Canada 179, 187, 212
Canada's Metis Uprising 114
Canadian National Exposition 201
Canadian River, North America 111, 139
Canavan, Greg 191
Canobi Station, Aust 8, 68
Canton, Frank 109, 110
Canyon de Chelly, USA 26
Cape Grenville, Qld 172
Cape Melville, Qld 124
Cape River, Aust. 4
Cape York Peninsula, Qld 16, 82, 121, 171
Capella, Qld 42, 97
Capertree Station, Aust 192
Capone, Al 14
Carboni, Raffaello 93
Carcoar, NSW 135
Carcoory Station, Aust 16
Cardew, Pollet 64
Cardwell, Qld 170
Carl, Herman 42
Carleton, General 26
Carlton Hills Station, Aust 69
Carnarvon Ranges, Qld 60, 66
Carpenter, Tommy 37
Carr, Second Lieutenant 67
Carriers Union, The (Aust) 96
Carrillo, Capt 45
Carrington, Colonel 23
Carson City, Nev 163
Carson, Colonel Kit 26
Carthage Female Academy (USA) 138
Casey Jones bucking horse 187
Casey, R.G. 99
Casper, Wyo 109
Casserly, Peter 100
Cassidy, Butch 219
Castle Hill, NSW 88
Castlemaine, Vic 144
Castles, Henry 54
Caton
   Bill 198
   Jim 197, 203
   Paddy 196,197
Catron, Tom 48
Cattle Creek, Aust 64
Cattle Trails of America 28
Cawood, Mr. Government Resident 75
Central Queensland Employers Association 96
Champ, horse 215
Champion Bronco Buster of the Plains 205
Champion, Nathan 'Nate' 109
Chapman, Caroline 127
Charleston, Ariz 44
Charleville, 'Qld 144
Charley, Aborigine 14
Charters Towers, Qld 41, 43, 170, 174, 196
Chatham Islands, NZ 58

Cherbourg Aboriginal Settlement, Qld 64
Cherry Creek (USA) 114
Chester Creek, Aust 4
Chester Range, Qld 124
Cheyenne Club, The (America) 108, 182
Cheyenne Frontier Days Rodeo 190, 200, 202
Cheyenne Indians, America 77, 79
Chicago, Ill 114, 158, 180
Chickasaway Valley, USA 21
Chiricuhua Apache American Indians 86
Chillagoe, Qld 42
Chinchilla, Qld 204
Chippewa, The, Riverboat 158
Chisholm Trail, USA 22
Chisum, John 47
Chivington, Col. John M. 77
Choate, Monroe 25
Christchurch, NZ 100
Chuckwagon 27
City of Melbourne, ship 171
Clancy, Frederick Melton 'Foghorn' 201, 202
Clandestine, stallion 39
Clanton
    Billy 44
    Ike 44
    N.H. 'Old Man' 44
    Phin 44
Clarence River, NSW 61
Clark, William 54
Cleary, Cammy 205
Clermont, Qld 96
Cleveland, President Grover 82
Clifford Downs Station, Aust 10
Clifton Station, Aust 39
Cloncurry River, Aust. 7, 68
Cloncurry, Qld 68, 69, 157, 174, 177, 193
Clum, John 86
Clump Point, Qld 170
Clumper, horses 150
Coachers, horses 143
Coal River (Newcastle) NSW 90
Cobb & Co Old Drivers Assn (Aust) 147
Cobb and Co (Aust) 142, 143, 145, 157, 162, 192
Cobb and Co Museum, Toowoomba, Qld 144
Cobb, Freeman 143, 160
Coburg Peninsular N.T. 148
Cochise County, USA 44
Cody, William Frederick 'Buffalo Bill' 134, 165
Cody, Wyo 84
Coen, Qld 122, 124 ,171, 175
Colborn, Everett 200
Coleman River, Aust 174, 177
Collarenebri, NSW 36
Colletts Creek, Aust 150
Collins Street, Melbourne, Vic 143
Colorado River, USA 25
Colorado Volunteers, 3rd 77

Columbia River, USA 158, 179
Columbia, riverboat 179
Columbus, Christopher 49
Comanche American Indians 26, 85
Combo waterhole, Qld 32
Comet Downs Station, Aust 123
Comet River, Aust. 69
Conargo, NSW 213
Conargo Hotel, Aust 213
Concord coaches 162
Concord, N.Ham 162
Condamine, Qld 203
Conglomerate Range, Aust 82
Connell, John 98
Conniston Station, Aust. 74
Cook
    Captain James 49
    Jimmy 203
    Judge R.G. 138
Cooktown Independent newspaper (Aust) 126, 170
Cooktown, Qld 122, 131, 172, 175, 176
Coolgardie, W.A. 150, 157, 172
Corella Creek, Aust 34
Costello family 11
Cottonwood Ranch, Canada 201
Coutt, John 74
Cowan, Les 198
Cowling, Edward 103
Cox
    Matthew Dillon 5
    Mrs. Hannah 64
    Thomas Price 131
Coxon, Charles 66
Cradlebaugh, Federal Judge 107
Crane, Jim 46
Crawford, Ray 198
Crazy Horse (Tashunke Witko) American Indian 23, 81, 83
Crazy Woman Creek (America) 110
Crear, John 54
Croaker, 'Greenhide' Sam 6, 10
Croll Creek, Aust 122
Crombie, Andrew 96
Cromwell, Oliver 87
Crook, Gen. George 80,81,83,86
Crooked Mick .210
Crouch, John 57
Crow American Indians 81, 84
Crowfoot, American Indian 82
Croydon, Qld 171, 172, 177
Crush, Tom 134
Cruz, Florentino alias Indian Charlie 46
Cueing Pen 14
Cullin-La-Ringo Station, Aust 64, 67, 69
Cummings, Jim 34
Cunnamulla, Qld 38
Cunningham, Phillip 88
Curio, bucking horse 198, 200, 203
Custer, Lt.Col. George Armstrong 79, 81, 83, 136
Custer's Last Stand, America 23, 79, 83
Cuttaburra Branch Station, Aust 39

Cyclone Agnes (Aust) 156

D

D'Aguilar Range, Qld 168
Daddy, Aust.Aborigine 59
Daggs
    P.P. 112
    W.A. 112
Dagworth Station
    Gulf Country, Qld 155
    Kynoona, Qld 32, 102
    Woolshed, Qld 87, 101, 102
Daisy, The boat 168
Dajarra, Aust 31
Dalby, Qld 67
Dalgangal Station, Qld 168
Dallas, Tex 139
Daly River, Aust. 5
Daly Waters Station, Aust 6, 148
Dangar, Henry 58
Dargin, Arthur 192, 193
Dargin's Grey, horse 135, 185, 192, 195, 197, 200, 203
Dariilba, Aust.Aborigine 38
Darling Downs, Qld 39, 96, 100, 101
Darling River, Aust 159
Darlot, West.Aust. 173
Darwin, N.T. 5, 37, 131, 148
Davis, a convict 60
Davidson & Smith, saddle 191, 199
Dawson River, Aust 62, 69, 166
Dawson, Smokey 198
Day, Edward Denny 59
Dayboro, Qld 169
Deal, 'Pony' 44
Deer Trail, Colo 205
Delaware American Indians 85
Denver, Co 29, 78
Derby, W.Aust 16, 12
Derwent River, Tas 55
Descent, bucking horse 200
Devine, Edward'Cabbage Tree Ned' 145
Devoncourt Station, Aust 69
Dewes, Police Magistrate 92
Dexter Saloon, Alaska 46
Dharuk, Aust.Aborigines 52
Dick, James 175
Dickie
    Elizabeth nee Fasken 169, 171
    Grace nee McCready 170
    Grace, 'Betsy' 170, 172
    Henry 169
    John 'Stringybark Fox' 122, 169
Dillinger, John 114
Dixon, Father John 89
Dodge City Globe, Newspaper 48
Dodge City Times newspaper 137
Dodge City, USA 18, 22, 136
Dodger, Aust. Aborigine 75
Done Gone, bucking horse 200
Double Tough, bucking horse 200
Dougherty, Mick 145
Douglas, Major Sholto 57

Index

Doveton Gold Fields Commissioner 91
Dow, George 23
Doyle, John 123
Drayton, Qld 67
Dreaming, The Aust.Aboriginal 50
Droving plants 19, 30
Du Fresne, Marion 55
Duchess, Qld 157
Duffy, Bill 137
Dufran, Phil 110
Dulacca Station, Aust. 61
Duggan, Robert 88
Dull Knife, American Indian 81
Dumas, Alexandre 128
Dunback, NZ 146
Dundalli, Aust Aborigine 73
Dunedin, NZ 145
Dunn, Andrew 54
Durack family 11
Durack, Long Michael 11
Durham (shorthorn) cattle 18

E

Earl, Ned 195
Earp
  Bessie 45
  Jim 45
  Morgan 44
  Virgil 44
  Warren 46
  Wyatt 44, 137, 219
Eastwood, Clint 218
Easy Money, bucking horse 207
Eaton, Mr. 59
Ebagoola, Qld 175
Ebor Creek, Aust 97
Echuca, Aust 159
Edmonton, Qld 174
Edwards, Doug 199
Egan, Ted 116
Eglington, Sub Inspector 69
Eidsvold, Qld 167
Elder, Sir Thomas 34, 156
Eliza Jane, Riverboat 159
Ellenwood, Joe 114
Elliott
  Joe 109
  Verne 201
Ellsworth, Kans 22, 117
Elsey Creek, N.T. 6
Elthan, Kent, Eng. 51
Elvire Station, Aust 29
Emerald, Qld 67, 97
Enniscorthy, Ireland 88
Eora Aust. Aborigines 50, 79
Epslom, Private 54
Erath County, Tex 186
Ernshaw, Harry 45
Esk, Qld 168
Eskew, Col. Jim 201, 202
Estuarine Crocodiles 24
Etheridge River, Qld 156
Eucalyptus, W.A. 133
Eucla, W.A. 150

Eugene, Oreg 179
Eulo Queen, The 119
Eulo, Qld 120
Eureka Hotel, Aust 92
Eureka Stockade, Aust 87, 93
Eureka Flag, The 93
Eurombah Station, Aust 64, 166
Euston, Aust. 159
Evans, Jesse 48

F

Fairbairn
  Charles 98, 99
  George 95
  George Jn 95
Fair Cape, Qld 172
Faithfull party massacre 3
Faithfull, George 3
Fanny, Aborigine 14
Fargo, William G. 160
Farleigh Station, Aust 70
Farlow, Charlie 194
Farquharson
  Archie 12
  Harry 7, 12
  Hugh 12
Farrington, Mag. Mr. H.H. 135
Fausho Ballagh, bucking horse 204
Federation Hotel, Brocks Ck, N.T. 134
Ferguson, Lark alias Pete Spense 46
Festival of the Bunya fruits, Qld 60
Fetterman, Capt. William J. 23, 83
Finke River, Aust 14
First Fleet, The, Aust 49
Fischer, John 139
Fisher, C.B. .6
Fisher, Jim 196
Fitzpatrick, Paddy 7, 9
Fitzroy River, Aust 11
Five Minutes to Midnight (Old 5),
  bucking horse 200, 201, 202, 209
Flagg, 'Black Jack' 110
Flaxie, bucking horse 200
Flemming
  John 59
  Mick 35
Fletcher
  Curly 206
  Thomas 92
Flinders Island, Aust 57
Flinders River, Aust 4, 7
Flora Valley Station, Aust 10, 172
Floraville Station 8
Floyd, 'Pretty Boy' 114
Flying Devil, bucking horse 187
Follin, Noel 129
Foly, Ned 59
Forney, Tex USA 187
Forest Creek Diggings, Vic 143
Forestier Peninsula, Aust 57
Forrest, Alexander 10
Forsyth, Major George 78
Forsyth's Scouts, America 78
Fort Benton, USA 158

Fort Buford, USA 81, 82
Fort Dodge Old Soldiers Home (USA) 138
Fort Gibson, Okla 45
Fort Laramie, USA 23
Fort Larned, USA 77
Fort Leavenworth, USA 23
Fort Lyons, USA 77
Fort Marion, USA 86
Fort McLeod, Canada 179
Fort Phil Kearny, USA 23, 83
Fort Randall, USA 82
Fort Reno, USA 23
Fort Robinson, Neb 83
Fort Smith, Ark 140
Fort Sill, Okla 86
Fort Sumner, USA 26, 29
Fort Wallace, USA 78
Fort Worth, Tex 186, 201
Fossil Downs, W.Aust 11
Frame, Angus 204
Fraser
  Charlotte 63
  David 63
  Elizabeth 63
  James 63
  Jane 63
  John 63
  Martha 63
  Mary 63
  Sylvester 'West' 63, 66, 68, 166
  William 63, 66, 84, 166
Fraser's Revenge 66
Fraser Island, Qld 60
Frayne, Jack 35
Freedom on the Wallaby, poem 32
Freemantle, WA 173
French Prairie, Oreg 179
French, Jim 141
Frizell, Dr. 166
Fullarton River, Aust 70

G

Gabriel
  Bert 180
  Bill 180
Galena, Ill 180
Gallager, Colin 100
Gannister, Sax 54
Garah, NSW 184
Gard, Jack 'Speargrass' 155
Gard, William 155
Gardenshire, Emilnie 205
Gardiner
  John 3
  Mr. W. 192
Garraway, Roland Walter 124
Garretson, Fannie 136
Gaunt, C.E. 132
Gawler Showgrounds, SA 204
Gayiri Aust Aborigines 64
Geelong, Vic 94, 128, 145
Gentleman Cadets, The 99
Georgetown, Qld 152

Georgina River, Aust 4
Geronimo, American Indian 85
Ghost Dance, religion, America 82
Gibson Desert, Aust 173
Gilbert River, Aust 177
Gilbert, Maria Dolores Eliza 'Lola Montez' 127
Gilbertson, Ross 109
Gill Bros. Stock Contractors (Aust) 187
Gill
   Brian 199
   Jack Snr 204
   Kitty 207
Gilligan, 'Bumble' 72
Gipps, Governor George 58
Gippsland, Vic 117
Girty, Simon 85
Gladstone, Qld 61
Gledswood Station, Aust 177
Glen Innes, NSW 145
Glencoe Station, Aust 5, 6, 9
Glenelg, Lord, Secretary for the Colonies 58
Godiva, Lady 117
Goliad, Tex 24
Good Intent, ship 131
Goodenough, Henry 123
Goodfellow, Dick 123
Goodnight Loving trail, USA 29
Goodnight, Charles 18, 21, 24, 26, 47, 111, 219
Goolwah, S.A. 159
Goondiwindi, Qld 197
Gordon
   Adam Lindsay 184
   Catherine 5
   Hugh 5, 6, 7, 8, 29
   Wattie 5, 7, 29
   Willie 5, 6, 7
Goulburn, Major Frederick, Colonial Secretary 53
Goulburn, NSW 10, 11, 144
Gowrie Junction, Qld 203
Grafton, NSW 197
Graham
   Billy 111
   John 111
   Tom 111
Granada Station, Aust 71
Grand Hotel, Mackay, Qld 125
Grand River, USA 81, 82
Grand Island, Calif 178
Grant, President Ulysses S. 82
Grass Valley, USA 127
Gravener, Buddy 199
Great Cattle Drives, America 1
Great Victoria Desert, Aust 173
Greely, Horace 163
Greenough
   Alice 207
   Turk 202
Gregory River, Aust 4
Grey Billie, pony 184
Grey, Victor Herbert 121
Griffith, Premier Sir Samuel 97
Gros Ventre, American Indians 83

Grose, Francis 52
Grover, Sharp 78
Guadalupe Canyon, USA/Mex 45
Guadalupe River, USA 85
Gubberamunda Station, Aust 64, 67
Gubbie-Gubbie, Aust Aborigines 74
Guion Point, Aust 37
Gulf of Carpentaria, Aust. 7, 68, 131
Gumbardo Station, Aust 98
Gun Carriage Island, Aust 57
Gungabula Aust Aborigines 60
Gunn, Mrs. Aenas 154
Gunnedah, NSW 195
Gunshannon, William 56
Gympie gold field, Qld 122
Gympie, Qld 74

H

H.M.S. Reliance, ship 51
Hagerty, Kitty 118
Haight, J.C. 106
Hall,Walter R. 143
Halls Creek, W.A. 172
Hamilton gold field, Aust 175, 176
Hamilton Reefing field, Aust 175
Hamley & Co. 190
Hamleys Circle H.Brand, saddlery 190
Hamden, NZ 147
Hand, Dora 'Fannie Keenan' 136
Handlesack, Trooper 89
Handly, Tom 196
Hands, Albert 149
Happy Days Mine, USA 47
Hardcastle, John McDonall 126
Harding, Judge George Rogers 97, 99
Harmstons American and Continental Cirque, show 185
Hart, William S. 47
Haslam, 'Pony Bob' 166
Hastings River, NSW 72
Haun's Hill Massacre, The 105
Hawdon, Joseph 1, 31
Hawkesbury River, NSW 88
Hawkesbury Show, NSW 193
Hawksbury, battle of the 52
Hawkins, James 59
Hay, NSW 144
Hayes, Bully 130
Haynes
   Fanny 134
   Harry 134
Hays, Timothy 93
Hays City, Col 78, 136
Head, Harry 46
Hector, bucking horse 199
Helidon, Qld 73
Hell, Highwater & Hard Cases, book 204
Hells Angel, bucking horse 200
Hells Gates, Qld 124
Henderson, Prairie Rose .207
Henly, Cecil 201
Henry, Ernest 69
Henshke, Alan 199

Hepburn, Captain 3
Herberton, Qld 171
Heywood, Calif 187
Hickidy Waterhole, N.T. 12
Hickman, Maj.Ben 135
Hickman, William 06
High Back Bull, American Indian 83
Higbee, Bishop 106
Hirschberg, Joe 144
Hobart, Tas 5
Hobbs, William 59
Hodges, Percival Thomas 126
Hodgkinson goldfield, Qld 130, 170, 175
Hoffmeister, Samuel 'Frenchy' 32, 102
Holbrook, Ariz 114
Holden
   Dally 198
   Nora 207
Hole in the Wall, The (America) 107
Holladay, Ben – the Stagecoach King (America) 161
Holliday, John 'Doc' 44
Holt, Mary Ellen 155
Hoofs and Horns Magazine 198
Horn, Tom 86
Hornetbank Station, Aust 63, 69, 166
Horse tailer 30
Horse wrangler 1, 30
Horsehead route, USA 26
Hotham, Gov. Sir Charles 92
Houck, James. D. 113
Howe, Jackie 100
Howlong Station, Vict. 3
Hoyt, Milo 141
Hugh River, Aust 14
Hughenden, Qld 4, 5, 69
Hull River, Aust 126
Hume, J.B. 164
Hunkpapa, American Sioux Indian 81
Hunt
   Elizabeth Jessie 'the Lady Bushranger' 135
   James 135
   Susan 135
Hunter, Governor John 52
Hunter, Robert D. 48
Huonfels Station, Qld 155

I

Idriess, Ion 173, 175
Iliff, John Wesly 29
Illawarra Station, Aust 35
Independence, Mo 158
Indian Charlie alias Florentino Cruz 46
Indian Hater, The 85
Innisfail, Qld 177
Inverell, NSW 12, 34, 58, 145
Inverway Station, N.T. 12, 13
Ipswich, Qld 63, 67, 168
Irish Revolution 88
Ironhurst Station, Qld 155
Islands Station, Qld 155

Index

Isle of Skye, UK 10
Ivory, Buster 187
Iwo Jima Hill 200

J

J.A.Ranch, America 18
Jackson, Pres. Andrew 77
Jacobs, Ed 113
James
    Frank 139
    Jesse 139
    Will 205
Jane E.Faulkenberg, ship 129
Jardine
    Alex 16
    brothers 16
    Frank 16
Jayhawkers 22
Jeremiah Johnson, film 84
Jiman Aust Aborigines 62, 166
Jingle Bob Ranch, New Mex 47
Johns, Trooper J.R. 38
Johnson County War, The (USA) 107
Johnson County, Wyo 107
Johnson
    Jack 'Turkey Creek' 46
    John 54
    John 'Liver Eating' 84
    Maj.George 89
    William 88
Johnstone, John (Negro) 59
Jones
    Ben 110
    John Decy 'Galloping' 40
    Mr 192
Joseph, John 94
Juandah Creek, Aust 66
Juandah Station, Qld 66
July, Jim 141

K

K.C.Roan, bucking horse 200
K.C.Ranch, USA 109
Kaiabara Aust.Aborigines 60
Kalgoorlie, W.A. 150, 172
Kalkadoon Aust.Aborigines 68, 76
Kamilaroi Aust.Aborigines 60
Kamm, Jacob 158
Kangaroo Point, Qld 73
Kansas, USA 22, 23, 77, 117, 160
Kansas/Pacific Railroad 22, 78
Kapperamanna, S.Aust 14
Kapunda, S.A. 204
Katherine, N.T. 5, 6, 9, 154
Keane, Timothy 38
Keenan, Fannie 136
Kelly
    Benjamin 118
    James H. 'Dog' 136
    Ned 207
Kempsy, NSW 5
Kendrick, Idaho 190
Kenedy
    Jim 'Spike' 137
    Mifflin 137
Kenmuir farm, Aust 10
Kennedy, Alexander 69, 76
Kenniff, Paddy 219
Kennington, Capt. James 83
Keriengobeldi Station. Aust 59
Kicking Bear, American Indian 82
Kidman, Sir Sidney 29, 144, 157, 204, 219
Killalpaninna, S.Aust 14
Kilmeister, Charlie 58
Kimberley Region, W.Aust 18
Kings Cross, bucking horse 187
King, Governor Philip Gidly 52, 89
King of the Ring, Bucking horse 203
King Ranch,Tex 137
King, Richard 137
King River, Aust 38, 154
Kingaroy, Qld 185
Kiowa American Indians 85
Kirby, W.H. 36
Kirk, Edward 14
Kitchener, Lord 195
Klondike gold rush, USA 46
Knickerbockerbuckeroo, bucking horse 187
Knight, Pete 201
Knights Landing, Calif 178
Kurrajong, NSW 184
Kwiamble Aust.Aborigine 59
Kynuna, Qld 32, 101

L

La Perouse, Comte de 49
La Trobe, Governor Charles Joseph 91
Ladies Buckjump Comp. 207
Lady Augusta, Riverboat 159
Lady Bushranger, The (Aust) 135
Lafayette, ship 119
Lake Frome, Aust 33
Lake Mary, N.T. 4
Lake Michigan, USA 158
Lakeland
    Claudie 121
    Elizabeth 122
    Esther nee Culton 122, 123
    Iris Stanley William 123
    James 122
    Leo Percy Bruno 124
    William 'Battling Billy' 122
Lalor, Peter 92, 94
Lamb
    Bill 14
    Jem 59
Lamber, John 143
Lambing Flat Goldfield (Aust) 104
Lambing Flat Riots (Aust) 103

Lander River, Aust 74
Landsborough, William 4, 5, 33, 68
Lanny, William Aust. Aborigine 58
Las Vegas, N.Mex 48
Laura, Qld 172
Laura River, Aust 122
Lawrence, Kan 139
Lawson
    Henry 32, 175
    William 53
Laycock, Quartermaster 70
Lea, Captain 83
Lee, Bishop John D. 106, 107
Left Handed Bally, Aust.Aborigine 63, 66
Legislative Assembly for Barcoo 96
Legislative Assembly of Victoria 94
Leichhardt River, Aust 4
Leichhardt, Ludwig 5, 172
Leichhardt's Bar, Aust 132
Lennons Circus (Aust) 196
Leofric, Earl of Mercia 117
Leonard, Billy 46
Leviathan, The (Aust) 145
Lewis Springs, Ariz .44
Lewis
    Jim 148
    John 148
Limerick County, Ireland 127
Limestone Aborigines, Aust 74
Limestone Hill (Ipswich) Qld 63
Limit of Settlement Ruling 77
Limmin River, Aust 9, 131
Lincoln County war, USA 44, 114
Linklater, Billy 36
Liszt, Franz 128
Lithgow, NSW 192, 195
Little Big Horn River, USA 79
Little Big Man, American Indian 83
Little Raven, American Indian 81
Lloyd, Tom 205
Logan Downs Station, Aust 96
Long Tom, bucking horse 187
Longreach, Qld 98, 117
Lorne Station, Aust 98
Los Angeles, USA 47, 139
Louisiana, USA 25
Loving, Oliver 21, 26
Loyola, James 103
Lucanus, August 104, 130
Ludwig, King of Bavaria 127
Lukin, George 105
Lunatic, horse 167
Lutterill, E. 53
Lyons, Maurice 6
Lytton, Phillip 135

M

Macarthur River (Aust) 130, 131
Macarthur, John 88, 90
Macumba Station, Aust 198
Mackay, Qld 3, 169, 174, 204
MacNamara, 'Happy Jack' 99
MacPherson

Allan 61
Bob 102
Macquarie, Governor Lachlan 53
Madagascar 119
Madison Square Garden, N.Y. 209
Madrill, Mick 154
Magdalen Asylum for Destitute Women, New York 129
Maggie, ship 38
Mahon, Danny 203
Mahoni, Aust.Aborigine 71
Maidenwell, Qld 61
Mail Contract, Melbourne to Yass 3
Major Reno, Bucking horse 200
Major, Aust.Aborigine 75
Majors, Alexander 160
Man from Snowy River, movie 218
Mandrake, bucking horse 187
Maneroo Station, Aust 98
Manhattan Island, USA 77
Manifest Destiny 77
Manning River, Aust 72
Mannum, S.A. 159
Mantle, Harry 195
Manual for Shearers Disturbances, The 101
Maoris, NZ race 58
Marble Bar, WA 133
Marcus, Josephine Sarah 45
Maree, S.A. 157
Mareeba, Qld 156
Margaret River, Aust 11, 104
Marine Officers Association, The (Aust) 96
Marion Downs Station, Aust 31
Maritime Strike, The (Aust) 95, 96
Marrabel, S.A. 198
Marrakai Station 6
Martin
  Ben 34
  Jack 'The Orphan' 131
  James 'Shearblade' 99, 101, 103
  Tom 34
Martini's Rough Riders' show (Aust) 135
Mary Ann, Riverboat 159
Massachusetts USA 143
Masterson, Bat 137
Mataranka (Bitter Springs), Aust 15
Mateer, Billy 167, 168
Mauritius 119
Mayithakurti Aust Aborigines 68
Maytown, Qld 105, 122
McArdle,Bill 151
McArthur River, Aust 9
McCarty, Eddie 201
McConnell, Reason 164
McConville
  Doug 198
  Thorp 196, 198, 205
McCoy, Joseph Geating 22
McCready, Elizabeth nee Doak 171
McCready, Hugh Jnr 169
McDonald
  Charlie 11
  Donald 10, 11
  Willie 11, 29
McDonnell Ranges 14, 34
McGregor, James 40
McIlwraith, Sir Thomas 70
McIntire, John 51
McIntyre, Andrew 92
McJannett
  Bill 173
  Jim 133, 170, 175, 178, 192
McKegney, Michael 54
McKellar, Dorothea 167
McKenzie, Donald 11
McKinlay Ranges, Aust 70
McKinlay, John 68
McLaine, Ian 190
McLaughlin, Maj. James 82
McLaury
  Frank 44
  Tom 44
McLeay River, NSW 72
McLelland, James 130
McLeod
  Canada 201
  William 98
McMaster, Sherman 46
McNab, Jim 201
McPhee, Garry 187
McTaggart, Hilton 197, 203
Meekathara, W.A. 172
Melbourne Argus, newspaper 91
Melbourne Cup, horse race, Aust. 36
Melbourne Herald, newspaper 156
Melbourne, Vic 91, 128, 159
Metropole Hotel, Murchison WA 133
Metropolitan Hotel, Eulo, Qld 120
Mexico City, Mex 179
Miami, Okla 186
Michalany, Jack 200
Middle Concho River, USA 26, 27
Middleton
  Harry 114
  Jim 140
Midnight, bucking horse 200, 201, 203
Milbong, Jemmy 72
Miles, Gen.Nelson A. 86
Milliron Ranch, USA 205
Milner, John 14, 15, 148
Milner, Ralph 14, 15
Milo Station, Aust 98
Minerva, Qld 97
Mingo American Indians 85
Minneconjou Sioux Indians 82
Misama Island, Papua 133
Mississippi River, USA 21, 24, 25, 77, 158, 160
Missouri River, USA 158, 160
Mistake Creek, Aust 71
Misty, horse 192
Mitchell Break saddle 190
Mitchell, Qld 68, 96
Mix, Tom 47
Mobile, USA 21, 25
Mochila, mail bag 161
Mogollon Range, The USA 111
Molloy, Peter 196
Monk, Hank 162
Mono County, Calif 206
Montana, USA 22, 79, 83, 187
Montez, Lola 127
Montgomery Wards Mail Order Store USA 215
Monymus, Aberdeenshire, Scot. 169
Moody, Terry 199
Moore, Lieutenant William 55
Moppy, Aust. Aborigine 73
Morant, Harry 'Breaker' 115, 192, 193, 205
Moree, NSW 184, 197
Morehead River, Qld 174
Moreton Bay Penal Colony, Qld 60
Moreton Bay, Qld 72
Moreton Island, Qld 60
Moriori race, NZ 58
Morisset, Commandant James Thomas 54
Morris, Dick 132
Morton, Tex 187
Mossman, A.F. 71
Mossman, Qld 125, 195
Mount Cook, Qld 176
Mount Pleasant, Qld 169
Mount Somers, NZ 100
Mountain Meadows Massacre, The (USA) 105
Mountain Meadows, Utah 106
Mourilyan, Qld 170
Mozee, Phoebe Annie Oakley 82, 121
Mt. Lola, Calif 127
Mt. Abundance Station, Aust 61
Mt. Cornish, Station 5, 33
Mt. Gambier, S.A. 184
Mt. Isa Rodeo, Qld 182, 187
Mt. Isa, Qld 68
Mt. Magnet, WA 133
Mt. Minn, N.T. 6
Muckadilla Creek, Aust 61
Mudgee, NSW 33, 54, 135
Muir
  Ernest John 'Sonny' 34
  Johanna 34
  John 34
Mulhall, Lucille 207
Mulligan, James Venture 122, 170
Multuggera, Aust. Aborigine 73
Mulvenon, Sheriff William 113
Mundubbera, Qld 64
Mungindi, NSW 196, 197
Murchison GoldFields, W.Aust. 10
Murchison River, Aust 29, 172
Murgon, Qld 64
Murphy, Francis 96
Murranji Track, N.T. 9, 12, 24, 30
Murranji Waterhole, N.T. 12, 13
Murrant, Edwin Henry 'Breaker Morant' 195
Murray River, Aust 3, 158
Murray, Mounted Const. William George 74
Murrumbidgee River Aust 3, 158
Muswellbrook, NSW 59
Mustang, horse 180, 187

Index

Muttaburra, Qld 98
My Country, Australian poem 167
Myall Creek Massacre, Aust. 58
Myall Creek Station, Aust 58

N

Nannine, W.A. 133
Narrandera, NSW 198
Natrona County, Wyo 109
Nauvoo, Ill 105
Neagle, James de Lacy 63
Neave, Alf 135
Neches River, USA 25
Nepean, NSW 89
Nevada, USA 46
Neville Creek, Aust 126
New England Tableland, NSW 5, 10, 12, 34, 58, 145
New Mexico, USA 26
New Orleans, riverboat 158
New Orleans, USA 21, 24, 25
New Romley, Ohio 80
New South Wales Corps 88
New South Wales Rum Corps 55
New York Tribune newspaper 163
Newcastle Bay, Aust 171
Newcastle Lagoon, NT 13
Newcastle Waters, N.T. 91, 12, 34
Newman, John 63
Newton, Kans 22, 29
Niagra Falls, USA 185
Nichol, Lieutenant Francis 63
Nicholson River, Aust 8
Nicholson, John 54
Nickavilla Station, Aust 195
Nighthawk, herder 30
Ninety Mile Beach, Vic 117
Nive Downs Station, Aust 98
Nogoa River, Aust 64, 69
Norfolk Island, (off coast of Aust) 54, 90
Norman River, Qld 177
Normanby, horse 133
Normanton, Qld 7, 8, 68, 69, 130
North Pine River, Qld 169
North Platte River, USA 165
Northern Territory Lightering Co (Aust) 133
Northern Territory Supreme Court 72
Nueces County, Tex 29, 137
Nullabor Plain, Aust 150
Nunyarra, WA 133

O

O.K. Corral, USA 44
Oakley, Annie 82, 121
Oamaru, NZ 147
Oates, James 59
Oddy, Joseph Smith 123
Ohio River, USA 86

Old Blue, lead steer 24
Old Colonial, bucking horse 187
Old Gulf Track N.T. 5, 11, 24, 130, 152, 172
Old Middleton Ranch, The (USA) 112
Oondooroo Station, Aust 101
Oracle, horse 168
Ord River, W.Aust 11, 29
Oregon Steam Navigation Co (USA) 158
Outlaw Trail, The (America) 107
Outram, NZ 206
Overland Mail Company, USA 143, 160
Overland Telegraph, Aust 5, 6, 9, 14, 15, 148, 150
OverlandTelegraph Co. America 161
Ovens River, Vic 3
Owen Springs Station, Aust 34
Owen, William Lambden 167
Owens, Sheriff Commodore Perry 46, 114
Oxley, John, Surveyor General 53
Oyster Bay Aust. Aborigines 56
Oyster Bay, Tas 58

P

Pacific Telegraph Co (America) 161
Packhorses 152
Paddy, Aboriginal tracker 74
Paine, John 112
Paiute (Digger) Indians, America 76, 107
Palliser, George 58
Palmer River goldfield, Qld 104, 122, 130, 170, 175
Palmer, Edward 68
Palmerston (Darwin) N.T. 5
Palmerston, Christie 122
Palmerston, NZ 146
Palmerston, ship 132
Palmyra Estate, Aust 169
Palo Duro – Dodge City trail, USA 18
Palo Duro Canyon (USA) 111
Pandi Pandi Station, Aust 16
Pangerang, Aust tribe 3
Papua New Guinea 133
Parker
  James 56
  Judge Isaac 140
Parkes, NSW 151
Parkhurst, Charlie 162
Parramatta, NSW 51, 88, 193
Parramatta, battle of 52
Parry, James 9
Pascoe River, Aust 172
Pastoralists Federal Council, The (Aust) 95, 100
Paterson
  'Banjo' 8, 32, 102
  Captain William 52
  Lt.Governor William 89
Patrick Plains (Singleton) Aust 60
Patterson, Mr. 15

Peak Downs Riot, The (Aust) 99, 103
Peak Downs Station, Aust 97
Peck, John Murray 143
Peckham, Henry Ventalia 'the Fizzer' 154
Pecos Bill 210
Pecos River, USA 47
Pemulwuy, Aust.Aborigine 50, 79
Pendelton, Oreg 190
Pepper Pot, horse 85
Percy, Mr. J.B. 167
Perdido Valley, USA 21
Permanent Indian Frontier 77
Persepolis, Persia 4
Peshawar, India 156
Peterson, J. 148
Petrie, Qld 168
Phillip, Governor 51
Philp goldfield, Aust 175
Phoenix, Ariz 113
Pierce, Johnny 198
Pigroot Creek, NZ 147
Pigroot Track, NZ 146
Pikedale Station, Qld 196
Pilbara region, Aust 167
Pilton, Qld 39
Pine Creek, N.T. 130
Pinkerton Detective Association (USA) 109
Piper, Edward 21
Pitta Pitta, Aust Aborigines 68
Pittsburgh, Penn 158
Placerville, Calif 163
Plattville, Colo 202
Pleasant Valley, USA 111
Pleuropneumonia 31
Pocock, Roger Ashwell 179
Poingdestre, Lyndon J.A. 171
Pole Sulky, (Cape Cart), horse drawn vehicle 143
Pony Express Service (America) 148, 161, 162, 165
Port Augusta, SA 148
Port Charles, NT 133
Port Darwin, N.T. 133, 148
Port Douglas, Qld 170
Port Phillip, Vict. 3, 55
Port Sorell, Tas 56
Port Stewart, Qld 124
Portland, NSW 135
Portland, Oreg 178, 200
Powder River, USA 109
Powell, James White 71
Powhatan Indians, America 77
Prairie Rose 117
Prendergast, Jack 196
Price, Col.Tom 95
Privett, Samual Thomas 'Booger Red' 186
Proserpine, Qld 193
Prospect, NSW 89
Pybus, Harry 14

Q

Quantrill, William Clarke 139
Quebec, Canada 206
Queensland Labourers Union 96
Queensland Legislative Assembly 68
Queensland Native Mounted Police
   54, 61, 62, 67, 124, 171
Queensland Shearers Union, The 96
Queensland Shearers Wars, The 95
Queenslander, bucking horse 187, 198
Quinkan, The Aust 82

R

Rader, Rev. M.A. 111
Ragged Thirteen, The (Aust) 130, 133
Raheen, Ireland 93
Rainworth Station, Aust 97
Rancho de Los Laureles, Tex 137
Randell, William 159
Rangatee Station, NZ 100
Rankleburn Station, NZ 207
Rapid Creek, USA 83
Ravenswood, Qld 174, 196
Rawsons Place, Sydney NSW 185
Ray, Nick 109
Red Buttes, Wyo 165
Red Cloud, American Indian 81, 83
Red Gum Ridge, Eukey, Qld 153
Red Lily Lagoon, Aust 15
Red Sash Gang, The (America) 107
Red Tomahawk, Sgt. American Indian 82
Red water fever 9
Rede, Chief Commissioner 93
Redford
   Harry 5, 33
   Robert 84
Reed Rocks p.131
Reed
   Black Jack 'Maori Jack' 130
   Henrietta 130
   Jim 139
   Simon 158
   Ed 139, 141
Reedy Creek, Qld 168
Reichstein, Leo 198
Remme, Louis 178
Reno, Major Marcus A. 80
Republic of Victoria, The 93
Republican River, USA 78
Retreat Station, Aust 36
Ricardo, Maj. Percy Ralph 97
Richardson
   James 119
   Mrs 29
   W. 29
Richmond Downs Station 7
Richmond, NSW 120
Richmond, Qld 4, 177
Ricky Land & Cattle Co. 206
Riley, Bryan 90
Ringo, Johny 44
Rio Grande River, USA 210
Rio Station, Aust 69
Ripon, Wisconsin 190

Risdon Cove, Tas 55
River Bend property, Aust 198
Robbery Under Arms, book 33
Roberts, Johnny 198
Robertson, Alexander 143
Robinson River, Aust 9
Robinson
   Alan 'Bull Tosser' 207
   George Augustas 56
   Isobel 'The Eulo Queen' 119
   William 119, 121
Rockhampton, Qld 3, 5, 16, 67, 97
Rocklands Station, NT 4, 16
Rockwell, Porter 106
Rocky Ned, bucking horse 181, 196, 197, 200, 201, 203, 204, 209
Rocky River, Aust 123
Roebourne, W.A. 166
Rogers, Roy 215
Roll-up Tree, Aust 103
Roma, Qld 33, 61, 67, 120, 171, 188
Roman Nose, American Indian 78
Ronan, Dennis 'Jim' 13, 34
Ronekeennarener, Aust. Aborigine 57
Rooseveldt, President Teddy 207
Roper Bar, Aust. 9, 11, 132
Roper River, Aust 132
Rose Bay (Parramatta) 51
Roseby, Rose 67
Rosewood Station, Aust 74
Ross, Capt. Charles 93
Rotorua, NZ 119
Rouse Hill, NSW 90
Rowell, Harry 187
Royal Canadian Mounted Police 179
Royal Hotel, Macarthur River, Aust 133
Royal Mail Hotel, Eulo, Qld 120
Rumley, Ohio 80
Rum Rebellion, Aust 90
Russell
   Jack 123
   Russell, John 58
   Majors & Waddell, freighters, USA 149, 160, 162, 165
   Russell, William H. 161
Rutherford, James 143
Ryan, Dennis 90
Ryan, James 98
Ryan, Paddy 201
Rylstone Court House, NSW 136

S

S.S.Menmuir, ship 133
Sabine River, USA 25
Sacremento River, USA 178
Sacremento Union, The, newspaper 178
Sacremento, Calif 158, 161, 178
Salt Lake City, Utah 106, 200
San Antonio, Tex 160
San Carlos, N.Mex 86
San Diego, USA 46, 160
San Francisco, Calif 21, 24, 119, 127, 158, 160, 164, 178

San Quentin Prison, USA 164
Sand Creek Massacre, USA 78
Sand Creek, USA 77
Sarcee Indian Reservation, Canada 202
Sarra, Chris 66
Sasquatch 212
Saudi Arabia 157
Saunders, W.D.H. 24, 25, 26
Saxby, Jack 75
Schieffelin, Ed 44
Schultz, Willie 196
Scobie, James 92
Scone, NSW 197
Scotch College, Melbourne 3
Scott
   Andrew 63, 66
   Gen. Hugh 86
   Sir Walter 116
Scurvy 'Barcoo Rot', illness 9
Searcy, Alfred 131, 132, 134
Sears & Robuck mail order store USA 215
Second Ponds Creek, Aust 89
Seekamp, Henry 128
Sefton Creek, Aust 122
Sefton, Robert William 'Bob' 122
Sellheim, Warden 105
Seven Hills, NSW 89
Sheaffe, Roger 69
Sheffield, Arthur H. 175
Sheridan, General Phillip 78, 80, 81
Sheridan, Wyo 180
Sherringham, Jack 131
Shirley
   Eliza 139
   John 139
   Myra Belle 138
Shonrock, Johnny 30
Shonsy, Mike 109
Shoshone American Indians 81
Shultes, Lydia Crigler 112
Siege of Dagworth, The (Aust) 101
Sierra Madre Mountains, USA 86
Sierra Nevada Mountains, USA 163
Silesa, College, England 195
Simpson
   Bruce 20, 30, 204
   Jeff 18, 20, 30
Singleton, NSW 135
Sioux, American Indians 23, 78, 79
Sitting Bull (Tatunka Yotanka)
   American Indian 81
Skuthorpe
   Lance Jnr 184
   Lance Snr 40, 135, 184, 186, 192, 195, 196, 198, 202, 204
Skyring, Zachariah Daniel Sparkes 74
Slick fork saddle 188
Sloan, Tom 195
Smallpox epidemic, Aust 51
Smith
   Athol 192
   Charlie 46
   Joseph 105
   Rose 207
   Tom 109

Index

Smuggler, ship 133
Smyth, Father 94
Snips, bucking horse 187
Snyder, W.H.Dudly 21, 24
Sofala, NSW 135
Softly, Private 54
Solomon, V.L. and Co 133
Somerset, N. Qld 16
Somerset Dam, Qld 169
Somerset, Henry Plantagenet 168
Somersetshire Regiment 2nd 54, 94
Son of Curio, horse 200
Sonora, Mex 45
Southport, N.T. 150
Southwestern Exposition and Fat Stock Show 201
Southwick, Jonnie 128
Soward, Walter 148
Spaniard, Jack 141
Speewah, The 210
Spense, Pete alias Lark Ferguson 46
Spicer, William 104
Spider, Aboriginal stockman 7
Spider Dance, The 127
Spinifex, bucking horse 187
Spotted Tail, American Indian 81, 83
Spring Grove Station, Aust 119
Springfield, Ill. 22
Springsure, Qld 67, 97
St. Clair, Gen. Arthur 85
St. George, Qld 6, 7, 96
St. Helena Prison, Qld 101
St. Joseph, Mo 161
St. Louis, Mo 158, 160
St. Valentine's Day Massacre (America) 114
Stafford, Randall 74
Standing Rock Reservation, USA 82
Stanley River, Aust 167
Stanly, Dick 200
Stanton, Jack 197, 203
Stanthorpe, Qld 153, 196
Star of the East, mine, Aust 133
Starlight, Captain 33
Starr
  Belle (Myra Belle Shirley) 138
  Sam 139
  Sam Jnr 139
Statesman ship 5
Steamboat, bucking horse 200
Steel, Mrs. May 29
Stephan, C.A. & Co, Aust 189
Stewart's Creek Prison, Qld 42
Stillwell, Jack 78
Stilwell, Frank 46
Stinson, John 111
Stock routes of Australia 17
Stockton, Calif 160
Stone, Fred 104
Story, Nelson 22, 23
Strathdarr Station, Aust 96
Strawberry Roan, The, song 206
Strickland, Hugh 200
Stringybark Fox, John Dickie 171
Stuart, John MacDouall 148
Stuckey, Harry 123
Sturt, Charles 158

Sugar Foot, American bandit 163
Sulieman Creek, Aust 69
Sulphur Springs Valley, Ariz .44
Surat, Qld 119, 144
Sutherland, George 3
Sutters Mill, Calif 5
Suttor Creek, Aust. 3
Swallow Creek Government Station, Aust 54
Swan Hill, Vic 159
Swanton, James 143
Sweetwater River, USA 165
Sydney Bulletin Newspaper 194
Sydney Morning Herald, newspaper (Aust) 128
Sydney, NSW 52, 118, 185, 193

T

Tall Bull, American Indian 78
Tam O Shanter, horse 172, 176
Tanami Desert, Aust 10
Tapanui, NZ 207
T.A.Ranch, America 110
Tashunke Witko (Crazy Horse) American Indian 23, 81, 83
Tasman Peninsula, Aust 57
Tasman, Abel 55
Tasmania, Aust. 54, 59, 118, 121
Tatunka Yotanka (Sitting Bull) American Indian 81
Taylor
  'Ointment' 145
  William 133
Tedbury, Aust. Aborigine 53, 56
Temerebee, Aust.Aborigine 38
Temple Bay, Qld 172
Tennant Creek, N.T. 14, 148
Tenterfield, NSW 34, 207
Terra Australis 49, 75
terra nullius, dictum of 50
Terrindillie, Aust.Aborigine 38
Terry, Gen Alfred H. 80, 81
Tewksbury
  Ed 112
  Jim 112
  John 112
  John D. 112
Texas, USA 22, 47, 87, 117
Texas Fever 22
Texas longhorns, cattle 1, 22, 47, 24, 26
Texas Panhandle, Tex 18
The Bucket Waterhole, N.T. 12, 13
The Peak Station, Aust. 14
The Black Line, Tas. 57
Thode, Earl 201
Thomas, Capt. John Wellsley 94
Thompson River, Aust 11, 98
Thompson
  'Yorky' John 14
  Fred 144

  R.R. 158
Thomson, Capt.William Campbell 171
Thornborough, Qld 155
Thungatti, Aust Aborigines 72
Thursday Island (off coast of Qld) 131, 193
Thylungra Station. Qld 11
Tieryboo Station, Aust 61
Tilghman, Bill 137
Tillyronach Farm, Scotland 169
Timber Creek, Aust 35
Timmins
  Billy 196, 203
  Wayne 'Stumpy' 197
Tindall, W. 192
Tinnenburra Station, Aust 38
Tipperary, bucking horse 200
Tisdale Ranch, Wyo 109
Tom Handley's Buckjump show (Aust) 196
Tombigbee Valley, USA 21
Tombstone Cowboys, rustling gang 45
Tombstone, Ariz 44
Tonto Basin War, The Ariz. 111
Toomey, Noel 199
Toonbaggee, NSW 52, 89
Toowoomba, Qld 39, 67, 203
Top Springs, N.T. 9, 12
Toronto, Canada 201
Touch the Cloud, American Indian 83
Toulouse, Charles 59
Townshend, Dick 35
Townsville, Qld .4, 42
Tozer, Horace 98, 100
Travers, W. 6
Tremebonenerp, Aust.Aborigine 57
Trigger, horse 215
Trimmier, T.J. 21
Trinity River, USA 25
Truan,Fritz 200
True to Label Genuine Wieneke saddle 188
Truganini, Aust.Aborigine 58
Trust and Agency Co. of Aust. Ltd. 96
Tumbling Mustard, bucking horse 202
Turner, Jeff, The Indian Hater 85
Turnoff Lagoon, Qld .8
Twelve Mile Camp, W.A. 172
Twobar Ranch, Wyo 200
Twofold Bay, NSW 118
Tyson, James 38

U

Uhr, Wentworth D'Arcy 5, 130, 172, 219
Undekerebina, Aust Aborigine 68
Undercut fork saddle 188
United Pastoralists Assoc. (Aust) 96, 101
United States Hotel, Ballarat, Vic 128
Unmack, Theodore 'Ten Bob Ted' 97
Urquhart, Sub Inspector Frederick Charles 70
Utah, USA 105

V

Van Dieman's Land 55, 118
Vaqueros, Mexican cowboy 45
Varieties Theatre (America) 136
Vavau Islands 119
Venus, ship 118
Verge, Sam 123
Vermillion, 'Texas' Jack 46
Vern, Frederick 93
Victoria Hotel, Eulo, Qld 120
Victoria Hotel, Mackay, Qld 170
Victoria River, N.T. 6, 12, 37, 154
Victoria River Downs, Station, Aust. 34, 154
Villa Pancho, USA 86
Villard, Henry 162
Vinegar Hill, Ireland .88
Vinegar Hill, NSW 89
Virginia, USA 77
Virginia City, Mo 23
Voyageurs, American fur traders 158

W

Wacol, Qld 68
Waddell, W.B. 161
Wade, Billy 157
Wagner, John 143
Waite, Billy 193, 195
Walbiri Aust Aborigines 75
Walbiri Land Rights Hearing, 1982 Aust 75
Walcha, NSW 10
Waler, horse 180
Walker
   Bill 109
   Frederick 'Filibuster' 61, 68
Wall, 'Strawberry' Red 201, 202
Walla Walla, Wash 158, 190
Wallace, Captain Bigfoot 85
Wallan, Qld 61
Wallerawang Station, NSW 192
Walpole, Edward 57
Waltzing Matilda, song 32, 102
Warburton, W.A. 173
Warby, Jack 9
Warditt, Aust.Aboriginal 38
Warialda, NSW 34, 196
Warramunga, Aborigine people 14, 15, 148
Warego, Qld 239
Warwick Rodeo, Qld 182, 187, 191, 203, 206, 207
Washington, President George 85
Watson
   Edgar A. 141
   Ella 'Cattle Kate' 108
   Jack 123
   Jim 122
Wave Hill Station 10

Wayne, John 218
Wayside House, Marble Bar, Aust. 133
We of the Never Never, book 154
Wealwandangie Station, Aust 69
Webb
   Henry 200
   Molly 186
Welford Station, Qld 13
Wellington, NSW 196
Wells Fargo 160, 162, 164
Wells, Henry 160
Welsh, Pete 201, 202
Wentworth, William Charles 53
Weraeria Aust.Aborigines 58
West Point, US Military Academy 89
West, Frank 141
Westbrook, Maine 73
Westerby, Henry 'Yorky' 92
Western Myth 210
Wexford, Ireland 5
Whaka Whaka Aust Aborigines 60
Whistling Annie, bucking horse 187
White Dick 200
Whitefield Indian Territory, USA 141
White Hills Station, Aust 71
White Horse, American Indian 78
White
   Bob 145
   C.F. 196
   Sub.Inspector 97
Whitmore, James 150
Whitney, William F. 143
Whyte-Melville, George 195
Wichita, Kans 22, 45
Wide Bay region, Qld 60
Wide Comb Dispute (Aust) 103
Wieneke, John J.'Jack' 188
Wild Australia, show 40, 135, 185
Wileemarin, Aust.Aborigine 51
Wilkinsons Station, Aust 167
William of Orange 87
William, McCulloch & Co 159
Williams
   Charlie 37
   R.M. 157, 199, 215
Willowburn, Toowoomba, Qld 121
Wills, Horatio 64
Wills Massacre, Qld 64, 67
Wilson, Alex 75
Windradyne, Aust. Aborigine 53
Windsor, NSW 89, 90, 196
Winter, Peter 207
Winton, Qld 101, 193
Wiradjuri Aborigines, Aust 53
Wise, Capt. H.C. 92
Wockia Aust.Aborigine 68
Wolcott, Maj.Frank E. 108
Wolf Belly, American Indian 79
Wolfgang, Andy 152
Wollemi Ranges, Aust 135
Wolston Park Hospital, Qld 68
Wooding, Jack 14
Woodriff, Capt 89
Woods
   Abel 195
   Alan 199, 204
   Jim 187

William 187
Woodville, Miss 26
Woogaroo Asylum, Qld 68
Wooloomooloo Bay, NSW 50
Workoboongo, Aust. Aborigines 68
Wovoka, American Indian 82
Wright, Pricilla 119
Wunumara, Aust Aborigines 68
Wyandotte American Indians 85
Wylds Station 54
Wyoming Stock Growers Association, The 108,109

X

Ximenes, Sub Inspector Maurice 92

Y

Yahagoug, W.A. 133
Yarra River, Vict. 3
Yavapai County, Ariz 113
Yellow Waterhole, N.T. 12, 13
Yemmerrawannie, Aust. Aborigine 51
Yeppoon, Qld 167
Yintayintan, Aust. Aborigine 59
Young
   Bonny 200
   Brigham 105
   Thomas 17
Young, NSW 104
Younger
   Bob 139
   Cole 139
   Jim 139
   Pearl 139, 141
Youngers Bend, USA 140
Yowie 212
Yreka, Calif 179
Yuggera, Aust. Aborigines 73
Yuleba, Qld 144

Z

Zig Zag Brewery, Lithgow, NSW 192
Zigenbine, Edna 30

About the Author

Jack Drake grew up on a sheep and cattle property near Palmerston in New Zealand's South Island and was fascinated by anything to do with frontier history from an early age. Jack and the education system did not have a lot in common parting company when he reached fifteen to the mutual satisfaction of both parties.

Horses were his abiding passion but pony club and show ring soon gave way to the excitement of the rodeo arena. It was rodeo that first brought him to Australia at age nineteen in 1969. After three years of knocking about, he returned to the Shaky Isles.

Twelve years later, Jack returned to Australia to pursue better opportunities in the horse industry that had become his career of choice. He worked as a farrier, saddler and horse breaker until becoming involved in tourism in the mid 1980s.

Working on Queensland's Gold Coast, he presented Outback Shows at theme parks which featured stock horse routines, whip cracking, sheep dog trials, sheep shearing and horse shoeing demonstrations plus bush poetry recitals.

In 1990 Jack in partnership with Stellamary Matheson, set up Red Gum Ridge Trail Rides at Eukey in the Stanthorpe area of south east Queensland, and for over ten years they conducted horseback adventures from one hour to two days duration for visitors to the popular wine growing region. Jack and Stella ran the business and married during that time.

He began writing poetry in the mid 1990s with some success winning Australian Bush Poet of the Year 2001 with a humorous piece called '*The Cattle Dog's Revenge*' about a larrikin blue cattle dog named '*Woody*'.

Old injuries from rodeo days caught up with Jack in 2001 making it impractical to keep looking after a large string of horses. He and Stella closed the trail rides down in mid 2001 and Jack then began earning a living as a performance poet.

He has released two C.D.s of his work which have both been finalists in the Australian Bush Laureate Awards, and in 2004 his first professionally published book *The Cattle Dog's Revenge* put out by Central Queensland University Press, took out the coveted Golden Gumleaf Award for Best Book of Original Verse at the Bush Laureate Awards in Tamworth.

The Wild West in Australia and America' is his first attempt at an historical work. It is the result of a lifetime's interest in comparative histories of the frontier times of both nations.

Jack and Stella live on their *Red Gum Ridge* property in Queensland's Granite Belt and travel extensively to festivals and other venues where Jack performs his poetry. They still keep a few horses and cattle and are both actively involved in historical research about early times in Australia, New Zealand and the Americas.

Jack Drake, 2002

Photo by Sue Moore, Ballandean.

Courtesy Jack Drake collection)

The Wild West in Australia and America by Jack Drake
ISBN 1876 780 66 5

The Australian Outback and the American Wild West were two of the last frontiers in the territorial conquest and expansions of the 19th century. These frontier territories were wild, lawless and extremely colourful.

The Outback vs The Wild West by Jack Drake
ISBN 1876 780 67 3

Famous pastoral explorers and drovers; poddy-dodgers and horse thieves; land-grabbers; colourful females on the frontier; Cobb and Co and puffing billies; rough riders and rodeos.

Colonel Lionel Rose by Trish Lonsdale
ISBN 1876 780 82 7

Colonel Rose AM OBE, after a very distinguished military career, solved the bovine pleuro-pneumonia problem and pioneered the live cattle export industry.

Kajirri: The Bush Missus by Alexa Simmons
ISBN 1876 780 75 4

In 1948 young Adelaide girl Lexie Simmons went to live on the great Victoria River Downs station in the remote outback of the Northern Territory. Before she knew it she was married to a head stockman & thrust into the role of 'bush missus'. It was sink or swim - she swam!

From the Gulf to God Knows Where by Marion Houldsworth
ISBN 1876 780 89 4

Features the life stories of pioneering bushies from the Gulf including Ann Brunner, Bob Forster, Bluey Ellis, Ethel Donnellan, Charlie and Pauline Rayment, Bill Petrie, Rob Whelan and more. Due: July 2006

Red Dust Rising: Ray Fryer of Urapunga by Marion Houldsworth
ISBN 1876 780 52 5

This is the inspiring story of a northern cattleman who built up the Urapunga cattle station from nothing. From the 1950s to the 1990s, he lived rough and worked hard. He worked closely with the tribal Aborigines, made Urapunga a dry station, coped with the many crocodiles in the Roper River, fought against cattle diseases, hunted buffalo, built himself a homestead and survived.

The Bush and the Never Never by Gerald Walsh
ISBN 1876 780 51 7

From Victoria River Downs to the Paroo, and from the Darling to the Murray, this book evokes the eccentric characters, the swagmen, the pioneers and the great bush entrepreneurs who built pastoral empires. It also evokes harrowing tales of babes lost in the bush, and historical treatments for snakebite.

On the Wallaby by Gerald Walsh
ISBN 1876 780 62 2

Continues to revisit and explore the largely neglected but important aspects of life in the Australian bush – the deeds of colourful pioneers, bizarre incidents and little known or forgotten facts about rural life.

Ten Thousand Campfires by Rex Ellis
ISBN 1876 780 65 7

Rex tells of his bush experiences leading safaris in the Outback over a period of more years than he cares to remember. But he also gives hilarious accounts of the down-to-earth tourist safaris that he led in Europe, Africa and India.

Mulga Madness by Rex Ellis
ISBN 1876 780 77 0

Rex Ellis made the first commercial tourism crossing of the Simpson Desert in 1971 and in 1974 was the only person to cross Lake Eyre by boat. He has been a jackaroo, a sheep station overseer and a safari guide from the Kimberley to the Simpson and the Gulf.

Victoria Downs by Mary Roberts
ISBN 1876 780 85 1

This is a book about a Merino Stud with a well deserved and very high reputation and this history is by implication an informal, social history of the merino industry in south-western Queensland.

Another Era by Helen Arnaboldi
ISBN 1876 780 73 8

This book is an inspiring account of the outback NSW and Queensland pioneers in merino sheep, the Mitchell family, from 1865 and especially of Jim Mitchell from 1897 to 1992.

Hauling the Loads by Malcolm J. Kennedy
ISBN 1876 780 63 0

This is the first book to explore in detail the vital roles played by our beasts of burden in the development of outback Australia. A comprehensive history of bullock and horse teams.